Praise for Expressive Origins

Finally! A high-powered, scientific insight into epigenetics. The focus is on the epigenome and how to adapt versus change our lifestyle to achieve an alignment with our beneficial genes. There are many examples of lasting, positive change. A truly liberating read. Personalized approaches to achieve long-lasting results in your endeavors.

Dr. Markus Wettstein, MD
Endocrinologist

EXPRESSIVE ORIGINS

TALES OF HOW TWO STRANDS OF DNA IMPACT HEALTH AND LONGEVITY

Rachelle Simpson Sweet Ph.D.

WELLNESS BOOK ENDEAVORS PUBLISHING

Published in the United States of America

Wellness Book Endeavors
c/o Authentic Endeavors Publishing
Clarks Summit, PA 18411
Book Interior and E-book Design by Amit Dey
Cover Design by Ambicionz

Expressive Origins
ISBN: 978-1-955668-52-1 (Paperback)
ISBN: 978-1-955668-53-8 (eBook)
LCCN: 2023903158

WELLNESS BOOK ENDEAVORS PUBLISHING

Disclaimer

Although the authors and publisher have made every effort to ensure that the information in this book was correct at press time, the authors and publisher do not assume and hereby disclaim any liability to any party for any loss, damage, or disruption caused by errors or omissions, whether such errors or omissions result from negligence, accident, or other cause.

Neither the authors nor the publisher assumes any responsibility or liability whatsoever on behalf of the consumer or the reader of this material. Any perceived slight of any individual or organization is purely unintentional.

This book contains information related to health care. The resources in this book are provided for informational purposes only and should not be used to replace the specialized training or professional judgment of a health care provider or mental health care professional. Neither the authors nor the publisher can be held responsible for using the information provided within this book.

Please always consult a trained professional before making any decisions regarding the treatment of yourself or others. If you know or suspect you have a health problem, it is recommended that you seek your physician's advice before embarking on any medical program or

Table of Contents

Dedication

To my husband, Jon,
and my parents, Terence and Theresa Simpson.

I thank you for your support and belief in me,
seeing what I don't see.

Acknowledgments

This book came from the divine. A message heard from deep within, pulling it forward to something beyond me. I am grateful that I listened and overcame. I'm grateful for the intuition overriding the fear.

My journey to this point was not easy. I spent many years in school working, researching, and studying. Despite some naysayers (there have been a couple who stand out) telling me it was not possible for me, I completed three college degrees and a post-doctoral fellowship. Along the journey, there were some brilliantly supportive teachers and professors. Each of them assisted me in reaching my potential. This became the base from which I have continued to fight, grow, and learn, thus shaping my abilities to overcome limitations and find what works best for me.

My understanding of genetics, epigenetics, and nutrigenomics grows with each passing day. But none of this would have started without my chance discovery of Apeiron Zoh. Dr. Dan Stickler, Mickra Hamilton, and all the support staff and coaches make learning and understanding epigenetics accessible. They provide amazing education, genetic test kits and supplements.

To Teresa Velardi, you started me on this whole journey. A simple inquiry led to a chapter that led to a book. To Peggy Willms for all the enthusiasm and support with writing, grammar, and punctuation. You have both been so encouraging in this process.

I am grateful for Liz Parrish, who was so wonderful to work with on the Foreword. I am grateful to each of my contributing authors who trusted me on this journey and contributed their knowledge and personal growth. Each of them supports consciousness expansion of health and wellness through their own practices, as I do mine.

This book is for my clients and the clients of all the contributing authors and coaches I have met with at rounds and whiteboards. They teach me as much as I teach them. Trusting us with their code to unlock, decipher, and interpret. Each of them is also on a journey of self-exploration, looking inwards to the truth. Now that you, the reader, have picked up this book to learn and grow, you are also continuing your journey of self-exploration, just like many of us have done with books.

Finally, to family, friends, and colleagues, who have been on this life journey with me, some were directly contributing, some were watching from afar, and some are no longer with us but still voices in my head. Each of you has contributed to my life in so many ways, altering a small part of my epigenome.

May we all aspire to B.E.S.T. living.™

FOREWORD

Welcome to Your Code

Your journey is unique! Your life story is forged from your genes and you want your adventure to be one of personal achievement and joy.

BioViva Science

As the world of computing and software development has risen over the last several decades, so has the public understanding of its associated jargon. We learned that "code" was the instrument and underlying structure that made everything possible, from the mainframe of our computers and the user interface to the apps we now use to order rides, meals, and holiday gifts. The code in computing essentially defines the algorithms that run on the hardware, create an outcome, and drive decisions making our lives easier or more difficult if there is a problem in the code. The human genetic code may be seen similarly, and its output becomes the epigenetic code (a 4-base-bit, 3-bit byte machine code).

To simplify learning, we need to break words into their parts. "Epi" means above, and "Genome" is the instructions that make up a living organism. So, you can think of the epigenome as what is happening on the top of the genome or the active code. This is the code that is currently running on your biological hardware. A gene is a hereditary unit that transcribes and translates the proteins that build you. These proteins are made when a gene is epigenetically active or turned on. Most of the human genome is similar within our species. These are the defining factors that create our sameness in human appearance. However, some genes vary, and those variations create our uniqueness.

Almost all the cells in your body have the same genes you were born with. Have you ever wondered why your nose looks different from your toes? The answer lies in the epigenome—the genes active in your toes vary from those active in your nose, giving them the various functions you enjoy. The variation in the gene activities gives your toes and nose different epigenetic codes. Because of this, you certainly don't walk on your nose or smell with your toes. For simplicity's sake, the genes turned on create proteins and regulate other genes; those turned off do not. You are simply a masterpiece of different symphonic codes in different cells of your body that creates the most amazing biological hardware in the world, you.

Genes can be broken down into subunits called A, C, G, and T, an acronym or abbreviation for the four types of bases found in a DNA molecule: adenine (A), cytosine (C), guanine (G), and thymine (T). These four characters are the ingredients of all living things. They play out in everything around you in the living world. Your genes come from your parents and are passed down through their heritage. Heritable traits are your eye color, height, and ability to tolerate food and drink. These traits are programs. Your skin color is also coded into the blueprint, as we see in vitiligo, a disease that causes the whitening of even the darkest skin. So epigenetics is at work all the time.

Looking at the natural world, you can see these programs running. Different combinations of these four ingredients are coded into every living species we know. From the color of a bird's feathers to the changing colors of the cuttlefish's skin, and finally to the hidden ability to see more colors or of heightened awareness seen in certain species. In real-time, epigenetics play out when a caterpillar changes into a butterfly or a grasshopper behaves like a locust. Not surprisingly, humans share many of their genes with other species, or at least genes so similar that animal studies can occur in one species that translate to human medicine.

If this all sounds like too much information, that is ok. My job is to introduce the idea and give you the basics to refer to if you want to learn more. I became interested in epigenetics in 2012 while working on a stem cell education project. I wondered why stem cells could regenerate tissues, but others did not. Now, I run a biotechnology company.

A word of warning. Wondering can lead to doing, and doing can lead to many amazing things you might not have planned for in your life. Those new things could change your health, your lifestyle, and what you do for a living for the better. It certainly did this for me. So, if you are naturally curious, then proceed. I am certain you have already benefited from your innate attributes. For the faint of heart or those who certainly don't have time for improvement, you may want to reconsider picking up this book.

Back to the stem cell. Stem cells have the same genes as other cells in our body, but a unique active epigenetic code defines them. The core difference in coding epigenetics is the difference between a cell that can divide into whatever cells are needed for a wound to heal, like a stem cell, or a cell that has undergone differentiation into a daughter cell with only one function, such as a skin cell. Almost all human cells have a limited number of cell divisions and, therefore, a limited lifespan. Again, most cells hold the same genes but express themselves differently. Consequently, that expression changes the outcome of the cell's life path.

Gene switching from on to off can and does change. Even aging can be read from the epigenetic code via an epigenetic mechanism called DNA methylation. In 2013, Steve Horvath released the first epigenetic clock that, over time, reads the changes in DNA methylation. This clock identifies differences between our biological age and our chronological age. Our biological age tells us how old our biology is compared with our peers of the same age, and our chronological age is how many birthdays or trips around the sun we have had. As the theory goes, an older biological age than a chronological one will lead to a shorter healthy life. This aging clock helps us target lifestyles and treatments that might help people live longer and healthier lives.

Then how do you change your epigenetics for a longer and healthier life? This book was written to explore the depths of these ideas. I can tell you one small and exciting fact about the future of epigenetic technology. In 2012, the Nobel Prize in Physiology or Medicine was awarded jointly to Sir John B. Gurdon and Shinya Yamanaka for the "discovery that mature cells can be reprogrammed to become pluripotent." This means that old cells were turned back in time and reprogrammed to be like the cell of a single-cell embryo. A single-cell embryo can grow into any tissue in the body and has immense regenerative ability. These are very exciting times we live in! And more importantly, since the discovery of cellular reprogramming, scientists worldwide have learned to harness reprogramming to avoid reversing to the embryonic state. This is important because a body needs cells that know how to differentiate the jobs cells do, or the body doesn't function. This is called partial reprogramming and can turn the cell's biological age back to a younger version of itself as measured by DNA methylation and other aging clocks without returning to the embryonic state.

Today, you can tinker with your epigenetics in beneficial and harmful ways. For example, if you drink a lot of alcohol, genes are expressed to try and protect your liver. The key word is "try." If you try intermittent fasting, genes that regulate metabolism will change. If you exercise, you get the upregulation of many beneficial genes that will make you look and feel better. Stress can prematurely age you, so don't worry too much about your epigenetics; try meditation and take holidays to benefit from your amazing algorithms.

Finally, I am delighted you picked up this book. Learning about epigenetics is learning about yourself and the world you live in. As you continue to read, you will learn how to test your genetics and DNA methylation. You can examine your unique genes for sleep, nutrition, athletics, hormones, cognition, or aging. Most importantly, you can understand what changes you could make in your lifestyle that could enhance or improve your future epigenetic outcomes. Together with a professional coach, you can further examine your unique set of gene expressions and how to implement changes based on your genetic blueprint to optimize your health and life span. You have already started the path to better health by being curious.

Elizabeth Parrish, MBA – CEO of BioViva USA Inc.

INTRODUCTION

How I Found Epigenetics and B.E.S.T. Living™

Living your best life is your most important journey in life.

Oprah Winfrey

Here I am, back at the gorge, ready to compare my physical and mental differences between my climb from two years ago. At that time, I was approximately 275 lbs. For the first time in years, I now weighed 190 lbs. - "One"derland.

Two years ago, the climb was slow, I was out of breath, and it was challenging to keep up with the others hiking with me. This time, I felt lighter, stronger, and more energetic. As I got to the gorge, I looked up the steps and began climbing. I thought it was easy! Initially, I wondered why it had been so difficult before. Quickly, I realized it was because of the extra weight I had carried. With my weight loss, I practically lost a small person. As I climbed, positive feelings rose. I felt liberated, and the excitement grew. I decided to run up the steps just before I reached the top. When I arrived, I turned to my fellow hikers and did the Rocky pose. Feeling the difference in my body was such a high point. As I reveled in it, I thought of all the new activities I would conquer and the heights I would reach.

Lying in bed the day after the hike, I brushed my hand across my abdomen and found a lump. This was curious; I had not felt this before. I told my husband I would check it out when we got home. The next day we boarded our return flight. During the flight, I became uncomfortable;

something was not right. The pain increased and became excruciating when I arrived at my house.

The next day I had an event. At the time, I was making and selling kombucha. As the event progressed, I was not feeling well, but I pushed through because I promised to be there. Then suddenly, I lost my vision, became woozy, and reached for my husband. I said I can't see, "something's wrong." There was a doctor at the event, and he and my husband rushed me to the clinic. He recommended that I have an ultrasound. Living about an hour or two from good medical care in Costa Rica, I went home to see if I could book an ultrasound in the capital of San Jose. In the meantime, I started to take some antibiotics to see if it was an infection causing my problems.

As the weekend progressed, I became more ill. The antibiotics were not working, and the fever increased. By Monday morning, I knew it was more serious. The journey to San Jose was painful. I felt every bump and rut. Finally, I reached the clinic, shivering and hunched over. I suspected it was a ruptured ovarian cyst because this had happened before landing me in the emergency room. Yet this was more intense, as if someone was wringing my insides out.

As the doctor took the ultrasound, he said, "You're in trouble. I need to get you into surgery right away." Lucky for me, one of the top cancer surgeons was in the clinic at the time and agreed to perform the surgery that evening. I didn't know then that the pain, fever, and tunnel vision were signaling that I was dying. My peritoneal cavity was full of toxic fluid. Not only had an ovarian cyst ruptured with a liter of toxic fluid, but there was a 2.5 lb. tumor wrapped around my ovary and was attached to my uterus and my colon. This explained why I had recently gained weight despite dieting and a lot of exercising. Unfortunately, there was no saving my female organs. The surgeon removed everything, throwing me into abrupt menopausal turmoil.

After the surgery, life was very difficult. I had to keep moving and be active to avoid complications from the surgery. But I became tired and sick very quickly. I wanted to go hiking and to the gym. I wanted that feeling of being at the top of the gorge. But my body was not responding the same way. Not only that, but my emotions were erratic. I was randomly tearful, and sometimes I thought I was losing my mind. My once good memory was thrown into chaos, and I couldn't even remember what seemed so easy for me before. I just wanted to go back to my "Rocky" moment. Despite lamenting about the past, I realized I needed to learn to live in the present moment and relearn life with this new body.

The follow-up gynecologist told me I was not a candidate for HRT (hormone replacement therapy) because the hormones could trigger another tumor growth. So, I began searching for natural remedies by looking for products that could help with the relentless hot flashes, calm the chaos in my mind, and stabilize my weight. I managed to find a couple of products that helped me through that time.

But something lingered. I had been given, in some regards, a second chance at life. This time I wanted to live differently. I had already been on a journey to improve my health for a couple of years, but now I wanted to extend my life and enjoy the world. I began to study the *Biology of Belief* with Bruce Lipton and *How to Become Supernatural* with Joe Dispenza. They introduced me to epigenetics. The idea is that the genetic code in our body, which we were born with, could be manipulated through our thoughts, beliefs, intake, and environment. As a result, we could actively do things to extend our life and live the most enjoyable, healthy life—reach a state of thriving.

Then by chance, I met a coach through Instagram. He was able to test my genetics and give me feedback on how I could work with my genetics to improve my life. He showed me a video featuring Dr. Stickler discussing the epigenetic possibilities of turning back time regarding biological aging. This was exciting to me, and I loved this new knowledge. Having a background in psychology and neuropsychology, I was always fascinated with how humans function physically and mentally. Now I wanted to learn how our body worked genetically, not just on the physical level but the extraordinary. However, getting my D.N.A. tested was not enough. I had to know more and eventually became a certified precision wellness practitioner in epigenetics. It opened a new world of experience, longevity, and optimization. I also found a new circle of friends and colleagues on a similar journey. Some are well-versed, and some are just beginning.

This newfound knowledge changed how I thought, breathed, ate, slept, and lived. I understood that a "normal" range might not be optimal for our blood work. The information we take in, thoughts, words, food, and chemical products all adjust our epigenetics or how our genes express. I learned some of the origins of my challenges, such as the "cookie jar" gene, the "sugar-tooth" gene, and "the fatso" gene. I have come to lovingly call these the fat trifecta. They have been with me my whole life. I inherited them, and my grandmother and father struggled with weight loss. While these are our origins (transgenerational genetics), I learned these expressions could be tweaked and adjusted like a dimmer switch. For example, refining what I ate or drank cleared my mind, and I began to think as I did in my 30s, despite menopause. My hot flashes decreased, and my deep restorative sleep improved. Understanding our baseline genetics allows us to hone our epigenetic expression and express our origins differently.

Now that I had learned to understand my genetics and epigenetic expression, I wanted to reach others and help them use this knowledge. So, I became a Certified Health, Well-being, and Lifestyle Medicine Coach to understand how to walk beside people with this knowledge as they journey to healthier and more optimized lives. Often my clients are very grateful because they understand that every person, including themselves, has a unique code. They realize they can now hone a path of experimentation to achieve the best version of themselves. Through my client's work and understanding of the science behind epigenetics, I created *B.E.S.T Living™*, a structure to follow to reach your ultimate expression.

This book, at its core, shows how we express our origins. Through the collaborative authors presented here and myself, you will learn you can express your origins differently. Like that dimmer switch, the expressive part on top of the gene, in some cases, can be modulated to express more or less to enhance well-being and, ultimately, a state of thriving. We are learning more and more every day how this can be achieved through research. Even now, whole genome sequencing has become more accessible and valuable in medical practice for treating heart disease, cancer, and neurodegenerative disease, which are among the leading cause of death. Shows like *Limitless* with Chris Hemsworth illustrate how genetics inform a complete lifestyle change simply because of awareness.

Awareness of our expressive origins allows undesirable effects to be tempered, and desirable effects can be enhanced. We are making these changes constantly, in every moment, so why not take the opportunity to make the most of what nature has given you?

B.E.S.T Living™ encompasses a few key elements, Breathe, Eat (Epigenetics, Energy), Sleep, and Thrive. Each contains within them other elements for growth and balance. These are fundamental building blocks to health, longevity, and overall well-being. This may seem simple, but sometimes in this chaotic world, simplicity is needed. Each element provides a block to build a better way of being. So, join the contributing authors and me on this epic journey.

Dr. Rachelle Simpson Sweet – *B.E.S.T Living™*

What is B.E.S.T Living™?

I choose to make the rest of my life the best of my life.

Louise Hay

B.E.S.T Living™ is how I conceptualized my health journey and my coaching program for clients. When I work with clients, I create a bespoke experience. Because everyone is epigenetically unique and requires a different starting point, each person can benefit from coaching in the four areas. Breathe, Eat (Epigenetics and Energy), Sleep, and Thrive. My job is to walk beside my clients on a journey of discovery about themselves. Each client finds their path in this process. Many are relieved that they hold the power to create the life they want instead of being directed on someone else's path. Often, they find it confusing. After years of following external ideas or someone else idea of healthy living, they come to understand that it's all within them. My job is to reflect and empower them and to hold a space of empathetic understanding. I have met many women who, like me, have struggled with their weight, bodies, and health since they were young. Each followed fad diets and exercise programs, leading to temporary relief but ultimately failure and humiliation. Many have developed metabolic conditions before working with me. Yet, this does not have to be. We hold a blueprint, a map, of how our bodies recover and thrive—a way to turn back the clock and reinvent ourselves.

There are four elements of B.E.S.T Living™. B: Breathe (Humans can only survive a few minutes without breathing.). E: Eating, epigenetics, and energy (Everything we do is affected by our epigenetics, and eating is one of the ways we can affect epigenome through nutrigenomics. In addition to nutrition, our energy through movement and resistance training balances and enhances epigenetic modulation). S: Sleep (To me, sleep is the foundation on which other areas of health sit; more vital than diet or exercise.). Finally, T: Thrive!

Each of these concepts is a building block to health and longevity. Depending on your goals, you may need to start in different places. Often sleep is the place I start first, which confuses clients who focus on losing weight. Each element can stand alone, be revisited, and be revised. It is often best to work on one aspect at a time, incorporating that element into your life until it is customary. We all know the difficulty in simultaneously optimizing our nutrition, exercise, and habits, which often leads to failure.

Our basic human instinct is to survive. Why just survive?

We have an inherent drive for self-improvement and growth. When we are exposed to positive scenarios, individuals grow, prosper, and are successful.

Abraham Maslow

Thriving refers to the state of flourishing or prospering. It can be applied to different aspects of life, such as physical health, mental well-being, and social and economic success. We can learn to thrive mentally, physically, and emotionally by becoming aware of who we are and where we want to go.

B.E.S.T Living™ is not a quick fix. It is a lifestyle—your unique, personal way of living. My journey, and the journey of many of my clients, was wrought with quick fixes that led to failure. At some point, I had to change my approach psychologically. In some ways, I had to change myself in small increments to become who I am today. I was the girl who hated physical education in school and has now become the woman who gets up at 6 AM to go to the gym.

Starting at age thirteen, I wanted nothing more than to lose weight. If you've read the story about my struggle with weight loss, you know I tried everything. The problem was that I continuously "tried to lose weight." Success started with a simple phrase, "I'm going to get healthy!" Simply reframing my approach made all the difference. Technology allowed me to understand things I'd never understood before, which led me on a journey to lose over 100 pounds, maintain that weight loss within a standard deviation, and begin to understand who I was at a multitude of levels and how to improve and thrive.

While I'm still on this journey with you, by no means do I have all the answers. However, I have a head start, understanding and having done significant research. The mission of B.E.S.T Living™ is to bring these elements to you, so you can understand who you are and begin the journey from "just survive" to thrive. While our learning journey is ongoing, my wish is that this book and B.E.S.T Living™ can lead you in amazement and wonder at how incredible you are and the amazing world we live in. My contributing authors and I present ideas, concepts, and client stories for entertainment, education, and enlightenment. That way, you can begin understanding how to change your life when you choose to do so, to look for the best methods and guides that will work for who you are at your core within your origins. These topics are just the tip of the iceberg. As we explore and expand, such a wealth of information is available. Science and consciousness are ever-expanding, revealing more to us than we could ever imagine. So much so that I couldn't possibly bring it all together in one book, yet I hope this information stimulates you to further investigate the idea of thriving and finding your ultimate expression of you.

Let's begin by learning how to breathe.

Breathe

Breathing in, I calm my body and mind. Breathing out, I smile. Dwelling in the present moment, I know this is the only moment.

Thich Nhat Hanh

I was in my late teens when I had my first panic attack. I felt my heart racing, struggling to catch my breath, and my vision went gray. The hyperventilation had caused my blood chemistry to alter, leading to these symptoms. What's crazy is that I hadn't even noticed it coming. It just seemed to come from nowhere.

If you've experienced a situation like this, you understand the feeling and how frightening it can be. No one likes the feeling of not being able to breathe. Without breath, there is no life. Most of us think it has to do with oxygen, that we're not getting enough oxygen to breathe. But that isn't the entire story. Normally, you breathe in oxygen and breathe out carbon dioxide. This is because the amount of carbon dioxide in your blood determines how much oxygen your body needs. But when you hyperventilate, like with increased anxiety, you breathe out more carbon dioxide than usual, and your blood levels drop, putting the system out of balance.

When I received my blood work, I could see this idea in action; one of the labs directly measures carbon dioxide levels in the blood. Carbon dioxide is a metabolic product of the cellular processes in the body that process lipids, carbohydrates, and proteins. My levels were lower, indicating that I had slight hypocapnia. What I've come to learn is that hypocapnia places my body at a metabolic disadvantage. Carbon dioxide plays various roles in the human body, including regulating blood pH, respiratory drive, and affinity of hemoglobin for oxygen. As with most things, having this sensitive system out of balance, such as too much oxygen and low carbon dioxide, can impact our metabolic system.

Until I read Patrick McKeown's book, *The Oxygen Advantage*, I had never considered the incredible impact oxygen and carbon dioxide had on my metabolic functioning. I always considered oxygen necessary for life, but as with everything, being off balance causes complications. More revealing was the idea that short periods of oxygen *reduction* could *improve* our blood oxygen-carrying capacity. This is done by increasing our sensitivity to our body's demand for carbon dioxide. In turn, it increases the maximum volume of oxygen (also known as VO2 Max) in our system available for our body to use. VO2 Max is the maximum amount of oxygen a person can use during intense exercise. If you're quickly out of breath while exercising (we all know this huffy feeling), it may indicate that you may need VO2 training with more intense exercise.

In genetic testing, we look at various types of VO2 capacities. The three categories of VO2 Max are cardiovascular, metabolic, and muscle. There is a vital genetic component to VO2 max; genetics can predict 25-50% of individual differences. I often find that clients have strong or elite genetic VO2 Max potential, but because of poor breathing habits, they are not maximizing these genes' potential.

According to Patrick McKeown, the quantity of air you breathe can transform your body, health, and performance. He explains that healthy breathing habits are just as important as healthy eating and engaging in physical activities. Unfortunately, breathing has been altered by chronic stress, sedentary lifestyles, unhealthy diets, and lack of exercise. These unhealthy breathing habits can lead to lethargy, increased body fat, sleep problems, respiratory conditions, or heart disease. As he explains, you may not even realize that the biggest obstacle to your health and fitness could be chronic over breathing.

As our lives become busier and our stress and anxiety levels increase, so have our breaths per minute. It's as if we are in constant "fight or flight" mode, which causes us to breathe more frequently. We have become chronic over-breathers. In addition, you'll see people breathing through their mouths rather than their noses. This increases oxygen levels but reduces carbon dioxide levels, which hampers our ability to carry oxygen in our blood. Oxygen is a key component for converting macronutrients into energy. There are three macronutrients: fat, protein, and carbohydrate, which we ingest from our nutrition, along with micronutrients, which are vitamins and minerals. With an adequate and efficient supply of oxygen, the conversion of nutrients into usable energy within the body works efficiently. Unfortunately, it is rare for the human metabolism to produce less carbon dioxide naturally. The main reason for our blood's low carbon dioxide levels, like hypocapnia, is increased exhalation, usually through our mouths. Often this happens unconsciously or while we are sleeping. Who has watched their relative asleep on the couch catching flies?

Some research suggests that mouth breathing may impact epigenetics, although the exact mechanism and extent of this impact are not well understood. Mouth breathing can lead to changes in the levels of certain molecules, such as nitric oxide, in the body, which can influence gene expression. Long-term effects of mouth breathing may be linked to chronic respiratory conditions and other health issues, like sleep apnea and asthma. However, more research is needed to fully understand the relationship between mouth breathing and epigenetics.

Within our blood, carbon dioxide is a type of electrolyte that helps control the amount of fluid we retain and the balance of acids and bases in our body. Of course, if we experience electrolyte imbalance can lead to confusion, weakness, and fatigue, just like a panic attack. This doesn't just affect the bloodstream; it also affects the brain. Maintaining carbon dioxide levels help our brain feel calm and relaxed because when we are in a state of hyperventilation, it typically means we are under stress or in danger, so the brain is on high alert. Over time, increasingly low carbon dioxide levels within the body can lead to a strong stress response and surges of adrenaline and cortisol.

We all know that stress and panic are bad for our cardiovascular system. We even joke about angry people; they "are going to stroke out." The cardiovascular system is also affected by carbon dioxide levels. Our arterial blood pressure and flow rate are subject to carbon dioxide's vasodilating (widening of blood vessels) effects. Carbon dioxide forces oxygen away from the blood to the muscles and organs so it can be utilized. This is called the Bohr effect. Optimum carbon dioxide levels and pressure keep blood vessels open, making them relaxed and smooth, which in turn helps blood flow. When blood vessels become more constricted, the heart naturally has to work harder to pump blood around the body, which can lead to high blood pressure.

Researchers indicate that normal breathing at rest should be about six to twelve breaths per minute or about half a liter of air per breath. However, many of us breathe 18 to 25 breaths a minute or up to a liter of air per breath. This over breathing is viewed as a low-grade form of hyperventilation which can lead to the imbalance of oxygen and carbon dioxide in our bodies. This low-grade hyperventilation signals to our physiology that we are stressed and in danger. As you will see later in the chapter, this can activate our sympathetic nervous system and put us in a state of constant "fight or flight." It is very difficult for your body to repair, heal, and regenerate in such a state. It won't even digest in this state, possibly throwing everything into storage (adipose tissue). When we are in this state of "just surviving," it is nearly impossible to work on being healthy. Because your brain thinks you have been literally fighting for your life, insulin levels rise rapidly, signaling you to replenish the calories. This is emotional eating at its core. Our many crazy attempts to be healthy will be thwarted by a body that thinks a wholly mammoth is coming through the door. But more on that later.

Dr. Joseph Mercola remarks that oxygen as a nutrient is often overlooked. Just like excessive calories can cause metabolic damage, excessive oxygen can prematurely damage our tissues by generating excess free radicals. He states that oxygen is very reactive when we over breathe by breathing in excessive oxygen and expelling too much carbon dioxide as our breathing rate increases. Chronic over-breathing leads to increased free oxygen radicals which cause oxidative stress. Oxidative stress is an imbalance between the systemic manifestation of reactive oxygen species and a biological system's ability to readily detoxify the reactive or repair the resulting damage. Simply, this oxidative stress leads to inflammation and can often accelerate the aging process. This is known as Inflammaging! Dr. Melissa will talk further about this in her chapter on longevity later in the book.

One of the ways that you can test your sensitivity to carbon dioxide is by measuring your BOLT (Blood Oxygen Level Test) score. According to William McArdle and colleagues, it is ideal if a person can hold their breath after normal exhalation for approximately 40 seconds. When I first measured my BOLT score, I was lucky if I got over 20 seconds. It drew attention to the relationship between my breathing, my health, my weight, and my performance in sports activities. I also noticed how much it affected my clients, many with hypocapnia, sleep apnea, and over-breathing.

Understanding the astounding effects of breath on my physiology, I now pay attention to my breathing, routinely measuring my BOLT score and my breaths per minute. I also assess this in my clients. We can improve our BOLT score and breathing rate by practicing beneficial breathing techniques. In Dr. Melissa's chapter, later in the book, she will describe a breathing exercise called box breathing that can also help calm your nervous system and decrease your breaths per minute. Also, in the book

Oxygen Advantage, you can find several other exercises that will help you improve your breathing rate during athletic performance and life. My favorite is the nose-unblocking exercise.

Throughout my research and discoveries, I've had some profound revelations through breathing techniques such as Heart Rate Variability (HRV) training, neurodynamic breathing, and meditation. These have directly reduced my stress and sense of panic. In addition, they have calmed my mind and cleared it for creativity and positive thought.

There is much research on the epigenetics of mindfulness, meditation, and mindfulness practices. Most of these practices have some emphasis on breath attention or breath awareness. Research has also shown the positive effects of meditation on the epigenome and how it regulates gene expression. In addition, many of these practices have been shown to alleviate stress-induced symptoms. Relieving stress symptoms lessens the "flight or fight" response. In turn, these will reduce the overall breathing rate.

Heart Rate Variability can be improved with deep breathing. I started hearing about HRV a couple of years ago, but until recently, I never really understood its impact. As more "wearables" began to give measures of "stress management" and "daily readiness," I noticed that HRV was a key component in calculating these biomarkers, along with sleep and movement. Then I was triggered to investigate further! In a meeting with other professionals discussing HRV levels and age, someone discussed a client in her 70s with an HRV of close to 100. My jaw dropped. Being significantly younger than this client but with an HRV SO much lower, I decided it was time to investigate this measurement further.

What is HRV, and why should we pay attention to it? HRV is the silent period in between heartbeats or pulses. Your heart doesn't beat at a steady rate. For example, if you have a heart rate of 60 beats per minute, your heart isn't beating every second consistently. Instead, there are slight millisecond variations between each heartbeat.

Proponents of HRV say it might be one of the best tools for tracking acute and chronic health concerns and represents an important mental and physical health biomarker of daily living and overall well-being. It has been indicated to be an effective marker for sleep, recovery, performance, and health of the heart and autonomic nervous system. Research shows that individual differences in HRV are associated with emotion regulation, psychopathology, cardiovascular health, and mortality.

While HRV manifests as a function of your heart rate as the variety of cardiac inter-beat intervals, HRV is essentially an interplay of the two branches of the autonomic nervous system, sympathetic (activating) and parasympathetic (deactivating).

The sympathetic system triggers the stress-induced "fight or flight" response I mentioned earlier. It raises your adrenaline and cortisol. The sympathetic system can be activated as a normal short-term response to stressful events. Yet, as we know, chronic stress and demanding life challenges can be seriously unhealthy for your mental and physical state, especially if you live in a highly sympathetic response state for long periods. When the sympathetic system is engaged, your HRV decreases because your heart generally beats faster and more consistently.

The parasympathetic system helps restore homeostasis (balance) in your body by regulating and lowering the response of the sympathetic system after an acute stress response. This system controls the rest, recovery, and digestion processes. After the stressful situation has passed, the parasympathetic system activates to slow your heart rate back down to resting. This increases your HRV to restore homeostasis, slowing your heart rate and beating less consistently.

A high HRV could indicate that your parasympathetic system is working and bringing calm and rest to your body. Conversely, a consistently low HRV could indicate that your sympathetic system is engaged, which means your body is stressed mentally or physically.

When moving to the epigenetic level, stress impacts our behavioral epigenetics. Although we can inherit stress-induced genetic expressions, we can also work to reverse this process. With certain lifestyle and environmental changes, we can reduce stress and reset our epigenetics for a healthier expression. For example, chronic and continued stress boosts cortisol and other glucocorticoids. These hormones impact histone coding and methylation of DNA. The histone code is a hypothesis that the reading of genetic information encoded in our DNA is, in part, regulated by chemical modifications (known as histone marks) along with similar modifications like methylation, a biological process by which methyl groups are added to the DNA molecule as part of the epigenetic code.

The increase in cortisol and other glucocorticoids can activate genes that express illness while deactivating healthy-suppressive genes. Too much cortisol and genes can "turn off," possibly leading to physical and mental health concerns. Research has shown that stress causes methylation and acetylation (introducing an acetyl group to the DNA) on various genes, especially neurological genes (those in the brain), where stress can also significantly affect genes that control memory and cognitive function.

To improve HRV, we can use various methods, such as time-restrictive eating (which I will discuss in the next section), reduction in alcohol consumption, reduction in body fat, meditation, exercise, and breathing exercises. When HRV training is combined with deep breathing, the synchronicity of heart rate and breathing increases the amplitude of heart rate oscillation leading to high levels of HRV. This explanation of HRV shows that the breath is integral to our heart functioning and overall stress reduction.

The first report linking HRV and breathing has been credited to Karl Ludwig as early as 1847. Research has shown that slower rhythmic breathing generally contributes to a healthier psychological and physiological response and stress reduction. During the research, it was observed that people who practiced deep breathing exercises for six months had more sympathetic nervous system inhibition and reduced their systolic and diastolic blood pressure. It was shown that deep breathing exercises reduced breathing rate typically to six breaths per minute while increasing the inspiratory flow rate. This increase in gas exchange reduced oxygen consumption and the heart's overall workload.

Earlier, I mentioned how people have switched to mouth rather than nose breathing. This affects our daytime performance and is an essential consideration during sleep. Breathing through the nose is how we were designed. Our nose is a natural filter and humidifier. It filters the air we breathe and warms it up, making it a perfect temperature for our lungs. It allows the amount of air necessary to balance oxygen and carbon dioxide. If our mouths are open at night, it could lead to increased metabolic changes, and in children, it has been reported to change the palate and jaw growth of the child.

One of the things that I have used myself is mouth taping, which seals your mouth at night and stops it from falling open. I sometimes use this technique with my clients if they do not have underlying breathing disorders. Some reported benefits of mouth taping are that it reduces snoring, lowers blood pressure, and strengthens immunity. Of course, if you feel you have more severe nocturnal breathing problems, you should ask your health provider for a sleep study to evaluate nighttime breathing. Mouth tape takes a while to get used to, and sometimes I can find myself mindlessly removing it in the night or finding it stuck to my sheets. But I persist with taping for the many reported health benefits.

In addition to HRV training and breathing exercise, I highly recommend a daily meditation practice. Research has suggested that meditation led to changes in gene expression. Studies have found that regular meditation practice is associated with changes in DNA methylation, which can affect the activity of certain genes. It was shown in a 2017 Harvard Medical School study that meditating just 15 minutes daily changed 172 genes that control inflammation, sleep-wake rhythms in the body, blood pressure, and how sugar is processed in the body. Many studies have shown that focused or mindful breathing reduces anxiety, depression, and pain.

Additionally, meditation has been shown to lead to changes in the expression of genes involved in stress response (fight or flight) and immune function. Over 300 studies in the past five years have shown a link between meditation and improvements for those with inflammation-related diseases like rheumatoid arthritis and inflammatory bowel disease. Not to mention Inflammaging!

You might feel a bit silly doing meditation the first time around, but it's worth the continued effort. Many people think it is about getting rid of all the thoughts in their heads. That is not true at all. But it will help quiet that neurotic maniac that follows you around all day, you know the one I am referring to. Meditation is about becoming aware and mindful of the external world and your inner dialogue. It is about noticing the place in your consciousness that offers calm and clarity and watching thoughts unravel.

Meditation helps you focus on mindful breathing, slows down your heart rate, enables you to relax, and can prepare you for a good night's sleep. I started with guided meditation because it helped reign in the distracting thoughts in my head— that negative super highway of ideas, worries, and self-conscious drivel gradually becoming more of a two-lane road with some positivity, gratitude, and loving-kindness sprinkled in. By meditating, I could begin to unravel the distracting thoughts and see them as less meaningful.

The time you make for mindfulness does not have to be long. Even five minutes will help ground you and improve your breath awareness.

There are several meditation apps available. Find one that works for you. I have used several along the way, including *Headspace* and *Gaia*. Currently, I use *Waking Up* by Sam Harris. It does not matter when you practice, just what you bring to the practice: your consciousness and its contents. However, it is crucial to start creating that space of open awareness and slow that super highway of thoughts down and, of course, your breathing rate.

I once had a profound experience that showed me how much psychology affects physiology regarding epigenetics. As a young child, I was at the beach swimming in the ocean, and a teenage boy thought it would be fun to grab my foot and pull me under. I still remember feeling the terror as I was pulled below the waves, and the bubbles of my breath escaped toward the surface as I sank. My mother and grandmother immediately came to my rescue. I surfaced, gasping for breath to find them chastising the boy. Such a small incident, you would think. But this, compounded with other events in childhood and adolescence, made a difference and impacted my epigenome, which altered my physiology.

Twenty-five years later, I experienced the power of regressive hypnosis while at a retreat. During the retreat, I revisited this incident and others. The idea is to visit the incident and reframe it to reduce the emotional charge of the event. One night following one of the sessions, I kept waking up. Every time I woke up, I said, I can breathe! There was a physical sensation of openness in my chest—a space to breathe like I had not experienced for as long as I could remember. The weight of psychological effects of these incidents literally weighed on my chest.

My individual epigenetics provided a way for environmental exposure to be "written" upon the epigenome, affecting the expression of the underlying gene due to environmental interactions. Within my body, the central pathway involved in response to the stressful event is the hypothalamic-pituitary-adrenal (HPA) axis. When a danger signal is recognized, central components (the amygdala, hypothalamus, and parts of the brainstem such as the locus coeruleus) of the central nervous system (CNS) will be activated to cause a stress response. Various neurotransmitters signal this transmission through the CNS. These then impact gene expression. Consequently, underlying gene variants may be activated or inactivated to cause differential health and stress-related outcomes through this HPA axis and regulate brain functioning.

Research on exposure to childhood traumatic events has been shown to interact with gene expression, possibly increasing the risk of psychiatric complications. Studies have shown that experiencing childhood trauma increases the methylation of certain genes. Also, brain-derived neurotrophic factor (BDNF) has been shown to play a role in childhood trauma and in association with mental health outcomes such as mood disorders. BDNF promotes the growth, differentiation, and survival of neurons in the brain. And is involved in neuroplasticity (the ability of the brain to form and reorganize synaptic connections, especially in response to learning and experiences). Structural brain changes are seen after traumatic events, and BDNF is hypothesized to be involved in these changes. Childhood traumatic events have been associated with decreased serum levels of BDNF and alterations in BDNF promoter methylation of the DNA.

The brain is a powerful thing. A cascade of neurochemicals changed my brain's operating system during that event and others like it. I became heightened for danger and attack, increasing my breathing rate and inducing anxiety and panic attacks. Then with guided intervention, I was able to "rewrite" this expression, changing how my brain responded because I had "neutralized" the event, transcribing this new response (danger is neutralized) on my epigenome. My experience with this reversal of symptomology is not uncommon. In Michael Pollan's book, *How to Change your Mind,* he witnessed people turning back their trauma with psychedelic-assisted therapy. Later, in Dr. Mickra Hamilton's chapter, she will touch more on this emerging work.

Ok, exhale! Breathe. You may never have thought about your breath as one of the most critical and integral factors in your state of health and well-being. Hopefully, you can continue examining your breath and consider it part of your overall well-being practice. Practicing a healthy way of breathing and reducing stress can lead to a healthier body and mind overall. This can lead to increased cardiovascular, metabolic, and mental health.

You can see that reducing breathing rates through practices such as meditation has been associated with epigenetic changes that may help reduce stress and improve overall health. This can decrease the production of stress hormones such as cortisol and help put the brakes on health problems, including inflammation, increased body fat, and metabolic disorders.

Enhancing and improving our breath allows us to increase our ability to become more open, aware, and mindful and engage in physical activities with more endurance and enjoyment. To live life while systematically improving and striving toward thriving.

Breathe Keynotes:

- Breathe through your nose at all times, even when exercising.
- Find a way to calm the "flight or fight" response with breath awareness.
- Find a place of open awareness through mindful practice.

Eat, Epigenetics, and Energy

Eat the energy you want to become.

Deanna Minich

We all eat; it is an integral part of our culture, ethnic group, social group, and sometimes our faith. It carries with it traditions, emotions, and beliefs, some of which date back to the genes of our ancestors. One of the most famous examples of this is the Dutch famine. When the offspring of women were followed over two generations, the grandchildren of women, undernourished during pregnancy, had increased adiposity (body fat). The deficiency in nutrition and lack of food modified the genetics of the grandchildren to have more body fat stores in case of future famine. This is a genetic adaption for survival. In addition, many studies have shown that prenatally and in childhood, nutritional cues affect our epigenome having beneficial or adverse effects in adulthood. Not just food but pollutants and toxins as well. By the same notion, our food and environment now affect our future grandchildren.

My great-grandparents lived through World War I, and my grandparents lived through World War II and other wars. During this time, there were food rations and shortages. Often, families didn't have enough food for all their children. I heard how my grandfather, one of nine children, got to eat the top of my great-grandfather's boiled egg, not the whole egg, just the top. My ancestors were working-class people; some lived in overpopulated houses in deplorable conditions by today's standards. What my great-grandmothers ate or didn't eat and the pollution they were exposed to was passed down through the generations to my grandmother, mother, and ultimately me.

In reviewing each of my parent's genetic data, they had one copy of the FTO (Fat mass and obesity-associated) gene, which was genotype GA (the specific genetic constitution of an individual allele). This gene is also known in the literature as the "fatso" gene. Through my conception, each passed one part of that genotype combination; in this case, A (Adenine) was passed to me. This created a "new" genotype of FTO AA in my genetic makeup. This genetic combination affects the gut hormone ghrelin and peptide YY which regulates food intake and appetite. This single nucleotide polymorphism (SNP), as it is known, can lead to increased food intake and may lead to weight gain and obesity as it ensures that I have body fat stores (energy) to "help me through times of food shortage."

What does all this mean? In theory, given that my ancestors experienced food shortages and I now live in a world full of palatable and exciting food, with a grocery store on every corner, I have struggled with losing excess body fat for most of my life. By the age of 10, I was 140 lbs. By the age of 13, I was on my first 'diet.' My weakness was always sweets. When I was a babe of 18 months, my great-grandmother introduced me to chocolate. What she didn't know was that I carried the "sugar tooth" gene (SLC2A) and the "cookie jar" gene (TAS2R38); each of these gives me the potential to overconsume sugary food. For those of us with these SNPs, once we get a taste for those rewarding foods, our brains won't settle for less. The GA variant of the "sugar tooth" gene makes me want to eat more sugar, and the AA variant of the "cookie jar" gene means that sugar deactivates inhibitions and reduces the "fullness" signal (satiety). Top that off with a bowl of AA "fatso" gene, and I had a recipe for lifelong increased body fat which led me to yo-yo dieting.

Globally, 43% of the population carries an A variant of the FTO gene, and 36% carry the A copy of the SLC2A gene. So, it is no surprise that over 40% of the US population has excess body fat. The FTO gene raises the risk of increased food intake through impaired central processing of fullness signals in the brain, particularly the hypothalamus. This was helpful for our ancestors, who may have gone days without food, but not so beneficial today. Research shows that those who carry the FTO AA genotype are 67% more likely to carry excess body fat. But it is more complex than this; genome-wide research has identified more than 500 areas in the genome that are associated with adiposity (body fatness). So, while I speak to those three genes as the "root" of my body fatness, it is much more complex than that.

Because genetics and human systems can be complex, it now makes sense why "dieting" fails for many people. It certainly failed me time and time again. You may have struggled with losing body fat for years if you're like many of my clients and me. Likely, what worked for someone else would not work for you because of our unique genetic variations. Healthcare professionals have told many of us to "eat less and exercise more" or "lose weight." Okay, easy for you to say! And so not helpful. This statement, in isolation, can further increase the shame and frustration I know I felt. And the feeling that you are failing in some way.

In reality, many of us have lost body fat, not just once, but repeatedly. That is not the hard part. The challenging part is keeping the body fat from coming back. This is because our brain, in some ways, is working against our desires. Not only do we have these genetic propensities, but also because we have a thermostat in the brain that monitors the body fat we carry on our bodies.

At the systemic level, a hormone called leptin acts to regulate hunger and energy balance. Leptin signals through its receptor in the hypothalamus to decrease food intake and increase energy expenditure. Loss of body fat can have a significant effect on leptin signaling. Leptin is produced by fat cells, and its levels in the blood are proportional to the amount of body fat the person has. When body fat decreases, leptin levels in the bloodstream decrease, which sends a signal to the brain indicating a negative energy balance, this results in an increase in hunger and a decrease in energy expenditure, leading to a reduction in the rate of fat loss. In other words, leptin acts as a satiety signal that helps to maintain energy balance and prevent excessive fat loss. In some individuals, the sensitivity of the hypothalamus to leptin may be reduced, leading to a condition known as leptin resistance. This can result in a failure of the hypothalamus to respond appropriately to the decrease in leptin levels that occurs with body fat loss, leading to persistent hunger and difficulty losing excess body fat.

In other words, once we reduce our energy intake (calories) to reduce our body fat, the leptin pathway senses this and works to correct the energy balance. Given the transgenerational genetics we discussed earlier, this system wants to prevent body fat loss (energy needed for that famine or food shortage). It does this by signaling the hypothalamus that leptin levels are low (because we are losing body fat). It thinks we are starving because it doesn't care how fat we are. It just wants to ensure we don't die from starvation; how ironic. Our brain then receives a message to ingest more food and reduce energy expenditure. So, you become hungry and tired. Your "diet" goes out the window. And the cycle of frustration continues. Then remember the stress hormones and "fight or flight" response we mentioned earlier. This only compounds our efforts to maintain less body fat. To top it off, we could adversely affect your resting metabolic rate (RMR) with each "diet" you engage in and discard.

The resting metabolic rate (RMR) is the total amount of energy you burn when your body is completely at rest. RMR accounts for up to 65% of our daily energy expenditure, physical activity can contribute upwards of 25%, and food thermogenesis (energy needed to digest) corresponds to approximately 10% of daily energy expenditure. Our lean body mass (LBM), the total weight of our body minus all the weight of our body fat, is usually regarded as the primary determinant of our RMR. When we eat more, our RMR increases and remains elevated after digestion, partially counteracting the effect of excessive energy (caloric) intake. Conversely, when we restrict energy intake (diet), our RMR is reduced. "Diets" may also cause both fat mass and lean muscle mass loss. When we lose lean muscle mass, we reduce our RMR further. Many forms of extreme "diets" do not preserve lean body mass, so when you start eating more and go off the "diet," it can lead to putting more fat mass back on the body, in part due to the lower RMR and part the leptin pathway mentioned above. The proverbial "battle of the bulge" continues.

For many of us, this has been a hopeless cycle. I lost "weight" in my 20s with hard work and exercise, reaching 155 lbs. before I got "diet fatigue." I vowed never to diet again, ultimately gaining the body fat back, and the cycle continued into my late 30s when I topped the scale at 297 lbs. I didn't think I ate poorly or "the wrong things," and I was exercising. But in some ways, this was a delusion, my mind playing tricks on me to maintain homeostasis.

In my teens, I remember my mother remarking to my grandmother, "She doesn't eat lots of fried food or cakes," to explain my higher body fat mass. She also remarked that I couldn't eat like my friends. Knowing what I now know, I think my genes played a massive role in these two statements. My genes (body) respond to the environment's signals (food, exercise, and other influences) based on my genetic predisposition, and my epigenome expresses those predispositions. So, it is true that I *can't* eat like everybody else, and certain foods affect me differently than others. I have my own genetic blueprint and epigenetic expression. What makes me gain or lose body fat is in my code, and it's in yours too.

This is the exciting part. The great thing about epigenetics and our amazing bodies is that each gene variant has a modification system. Those of us with the FTO variety AA can lower our risk of body fat gain by 27% by increasing physical activity. Also, our RMR is strongly influenced by exercise. So, if you are physically active, you can maintain a higher RMR, lean body mass, and less body fat than those who are sedentary.

Also, strength training, in particular, can help preserve and build our lean body mass while maintaining energy intake and may slow down the reduction of RMR as we age, regardless of our body composition.

Additional research has shown that increased body fat mass is associated with widespread changes in DNA methylation. The modification in DNA methylation was predominantly the consequence of excess body fat rather than the cause. These ever-changing methylation marks can be altered with the right lifestyle changes. Interventions such as physical activity and nutrition changes have been shown to change methylation patterns in different tissue types. The study of controls and successful body fat loss maintainers displayed similar methylation patterns relative to individuals with excess body fat. This suggests that reducing body fat mass can modify methylation marks and epigenetic expression. The bottom line is that our genes do not control us, and lifestyle changes can change how our genes express epigenetically.

What have I learned through my fat loss journey?

Everything changed for me one year when visiting friends. They were on the "Caveman Diet." It was entirely too much meat for me, and this was not a "diet" I would try or sustain. Given what I know now, I am convinced we should eat more like our grandparents and great-grandparents than a caveman. Also, I believe our ancestors were opportunistic and ate whatever could sustain them. Those cavemen did not sit around the fire, discussing what to have for dinner!

I digress. Anyway, they (my friends, not the cavemen) were making green shakes in their *Nutribullet* (a single-serve mini blender). Now, this I could do! I began changing my eating routines by following the recipes and eating suggestions that came with the machine. Feeling better, I began to lose body fat, about 25 lbs. overall. While the "diet fatigue" had not lifted, I transformed my mindset, starting with "I will never diet again" and "I am getting healthy." I realized I had approached fat loss incorrectly. It was time for a fundamental shift - stop "trying" to lose weight because, as I said, I had done that and failed too many times. Instead, I needed to embrace a way of life that would be sustainable and healthy. Understanding that it might not be perfect and wouldn't be a quick fix, the seed began germinating. Then by way of a gift, I got a wearable smart watch. This was a game-changer!

When I started my health journey, I used *Fitbit*. I now have a *Garmin*. It is an invaluable tool to help me stay on track. It was a revelation to see how much energy I was burning. For the first time in my life, I could see the energy I was consuming versus the energy I was expending. I wasn't blind in my eating plan, and this idea of "just eat less" came to life. It also allowed me to be flexible with my eating, as tracking my food intake was easy using the app on my phone. I could still have foods I enjoyed because I worked the calories into my daily energy limit.

I started with a 1000 energy deficit for a two-pound fat loss per week. Because I weighed nearly 300 lbs., I had a lot of leeway. How does this work? I burn energy throughout my day, working, eating, resting, exercising, and walking my 7,000 -10,000 steps. I could burn upwards of 3,000 calories per day or more when I was large, even more with increased physical activity. After all, I was carrying a whole other person around! Our activities vary from day to day. Therefore, I looked at my energy burned over an average week. By way of example, 21,000 calories per week. With a 1000-energy deficit, I could consume upwards of 2,000 calories daily. Living in this energy deficit works for me and allows me to be flexible with my food choices, especially if I burn extra energy. Following this idea, I started to see my body fat percentage diminish. You may have heard that wearables are often inaccurate, maybe up to 20%, but I can tell you that they are a great way to start, especially if you have a long way to go in your fat loss journey. If you have less to lose, you may have to dial in more closely and make more adjustments to optimize LBM (lean body mass).

These days, at a lower total body weight and LBM, I have to play with my energy and physical activities to reduce my body fat mass. I am not comfortable with a low energy deficit. After years of following plans, I have come to know myself pretty well. I know when I am on track and when I am off track. I also have to stay sane. I have discovered that exercising and movement are so important for me (and I kind of always knew it). Remember, those FTO genes are modified by activity. As you lose fat mass (adipose tissue), you naturally lose LMB, as adipose tissue contains some lean tissue. Also, your muscles are not required to carry that extra fat mass around, so they reduce to be more efficient. This lowers the RMR and, conversely, how much energy you burn. The bottom line, I was not able to eat as much as my "cookie jar" gene wanted me to. There is also a significant psychological component to this. Developing a healthy relationship with food and one's body is essential. Remember, I said this is a journey to improve your health in mind, spirit, body, and environment. You are with me on my journey. However, your journey will and should be different.

As I said, I learned that movement was so important to me. While there are many new, revolutionary ways to track your fitness and health, a quality wearable can make it super easy to get motivated, reach your step goal, watch food intake, or meet your energy burn goal by the end of the day. Some wearables may even come with the option to add friends where you can compete in daily or weekly challenges together for extra motivation!

One of the best and most valuable data a wearable can provide is your heart rate, especially when measuring your physical activity. By monitoring your heart rate data, you know when you reach your peak fat-burning zone. You will often hear people referring to Zone 2 physical activity. In Zone 2, your heart beats about 70% of your maximum heart rate (220 minus your age). So, if your maximum heart rate is 180, your Zone 2 heart rate would have a minimum threshold of 126 and a maximum threshold of 144; the more you can stay in that range without exceeding it, the better.

Moreover, Zone 2 exercise is vital to maintain healthy cardiac functioning. An emerging body of research shows that low to moderate doses of physical activity significantly reduce long-term risks of all-cause mortality. The optimal dose, or what has been termed the "Goldilocks Zone," for physical activity is at least 150 minutes per week of Zone 2 activity or 75 minutes or more per week of Zone 3. But not more than four to five cumulative hours per week of vigorous (heart-pounding, sweat-producing) exercise, especially for those over 45.

Wearables can be a great asset to becoming healthy. They provide useful data for your overall activity, internal health, and even sleep. You can better manage and reach your health goals because it keeps you accountable, especially if you don't get too obsessed with the data. Learning these key components can be a tremendous help with long-term success. You no longer need to second guess if your physical activity was in the zone, even with the percentage of error. You can also dial in your sleep routine, which is very important to body fat loss, which I will discuss in the next section. If the data itself stresses you out, you don't have to use these devices. There is more than one way to approach your goals, these were helpful for me, yet I understand we are all in different places.

When it comes to data, food tracking is something to think about. It was eye-opening when I started looking at labels and reading what was inside the package. I had never really paid attention before. But once I started tracking, I became aware of where I was out of balance. I became mindful of ingredients that I didn't want to ingest. And it's not just me. Many clients are shocked to discover this. I know that food logging can seem overwhelming or tedious, but many apps have made it easier by scanning bar codes to quickly add meals to your personal food library, helping you find them more easily the next time. Many restaurant dishes are in the data bank, too.

I logged my food daily for over a year when I started my health journey. By doing this, I became more aware of portion sizes and how much energy I was consuming. Weighing and measuring your food is recommended until you know your portion sizes. I may have taken it a little too far at times by taking my measuring cups to a friend's house who had prepared a meal for us. You don't have to be so crazy. Yet weighing and measuring may help you stay at an energy-burn level if you aim to lose or maintain body fat.

A study published in the Journal of the Academy of Nutrition and Dietetics found that individuals who kept a food diary lost significantly more weight than those who did not. Another study found that individuals who kept a food diary were more successful in maintaining overall weight over time. Some tools I found helpful along the way include the *Fitbit* food log, Chronometer, and the Carbon app. Again, you have more energy leeway when you need to lose more body fat; the less fat you want to lose, the more complex this becomes. Some clients are reluctant to logging. I suggest they try it for one month to learn what works and what doesn't for their individual eating goals. Then they can stop logging to determine how they do without logging and move to a more intuitive eating style. If the body fat loss slows down or stops, I recommend they resume logging to see where the leak may be. Again, assess your individual goals; this has worked for me so far. I have logged my food intake on and off throughout the years. Logging my food works best for me to maintain my overall goals. Also, as my eating habits change or I find new foods I like, I can see how these fit into my eating plan. It can be eye-opening!

Remember, the body fat may shift with other interventions, such as regulating stress, hormones, and/or sleep. We are complex beings; sometimes, we get in the weeds with a particular intervention. Life is an experiment. See what works for you.

Another way to gather data and log your food is with a Continuous Glucose Monitor (CGM). You can try this device on a trial basis or with a subscription. I completed a two-week trial and received tremendous insight. Given my experience, I recommend using it for at least four weeks. Learning about what was happening inside my body was amazing. As a result, I had to break up with Indian food or at least reduce my intake (bummer).

One of the things I learned through understanding my nutrition genetics was that I need more protein. I knew this from trial and error with "diets" over the years. At one point, I was almost entirely plant-based, with just a few eggs here and there. But this eating plan did not suit my unique genetic makeup, and I actually gained more body fat. Once I added more animal protein back into my plan, my body composition changed. Having the CGM on my arm confirmed this for me, too.

Dishes heavy in lentils, garbanzos, vegetables, and a little naan bread increased my glucose to the upper limits. This is okay if the glucose recovery period is normal, but mine was not. One particular meal I ate at around 7:30 PM – 8 PM. With no alcoholic beverages and no rice, my blood glucose stayed elevated beyond 1 AM, which, based on my reaction to other foods, even non-dairy ice cream, was not my norm. It got me thinking about my genetics and the type of food that work for me and against me.

Everything we eat is information to our body, and what we provide to our body in the way of that information can affect the outcome or our body composition. This was one of the biggest surprises of wearing a CMG. Over the years, I heard, "Don't eat that. It will spike your glucose," but it did not always ring true for me. In fact, with some experimentation, I found that if I consumed protein, such as eggs, before I ate food with a high glycemic index (a value assigned to foods based on how quickly and how high those foods cause increases in blood glucose levels), such as a banana, the "sharp rise" in glucose was nominal for me. I even did this with one of my favorite foods that I had previously given up, waffles. My glucose curve was "flatter" when I ate the eggs first and then the waffles.

Of course, I will not eat waffles regularly as they are refined, processed, and contain items I don't want to eat all the time—a choice I made for my B.E.S.T Living™ plan. The good news is that if I want to indulge in waffles once in a blue moon, it will not kill me or spike my glucose beyond my control, at least at this point in my health journey.

All of this can be incredibly comforting and calming to an anxious mind. It is powerful when we have knowledge. To paraphrase one of my clients, once she learned her genetics, she said, "I approach food differently now. I am not so rigid and feel more comfortable because I know what works best for me." That was music to my ears. After spending years worrying about everything that touches our lips, this knowledge gives us power, control of our choices, and solace.

Let's talk about macros! This subject often sparks a debate. I want to focus on macronutrients in relation to your specific genetics based on your specific genotypes. After all, it is not what percentage or amount "everybody" else says you should eat. It's what your genetics says. Analyzing my client's data has proven that most of them were eating all the wrong macros for their genetic propensity, so there is a good chance you are as well. In Laurie Kaplan's chapter, you will read an example of this with one of her clients.

Many of the crazy fads and trends have us eating at extremes leading to failure for most people. One thing is for sure, fat and carbohydrates will give us plenty of energy, but protein is the building block. Several factors, including the availability of amino acids, hormones, and physical activity, regulate muscle mass maintenance and function. Resistance exercise and protein consumption are two factors that have been shown to increase muscle protein synthesis and contribute to the maintenance of muscle mass. We require protein and its constituent amino acids to adequately facilitate the repair and build skeletal muscle cells in response to our physical activities.

Remember I said earlier that I had to break up with Indian food? Well, it wasn't just because of the prolonged glucose exposure. Eating mainly plant-based proteins did not provide *my* body with adequate protein for my genetic phenotype. Generically, I have an increased benefit from consuming protein in terms of appetite regulation, waist circumference, and body composition, with the strongest benefits epigenetically coming from 1.1 to 2.25 grams of protein for each kg of body weight. WOW! This was much more than I expected.

For example, if someone weighs 200 lbs., they could consume 100-200 grams of protein daily. Looking back at my food journals, I was only consuming around 80 grams a day which was significantly less than *my* personal requirements. My challenge was balancing my total energy goal while meeting the minimum protein goal for optimal epigenetic expression. To do this, I had to change some of my protein sources. To be on the lower end, I had to ensure I consumed around 30 grams of protein per meal and a 10-gram snack to meet a beneficial intake. Clearly, in the past, I was not working with my body to maintain adequate lean muscle mass to set myself up for vitality as I aged.

Unfortunately, the protein content of the adult body diminishes with age. Non-muscle mass (body fat) changes very little with age, whereas LBM diminishes extensively. Sometimes there will be an increase in body fat to compensate for these changes to LBM. Therefore, as we age, it is essential to get adequate high-quality protein. Not to mention, diet-induced thermogenesis is greater on high-protein eating plans than on high-fat eating plans. Meaning you are expending more energy daily just with digestion. Protein is also more satiating and reduces hunger pangs.

There are nine amino acids—histidine, isoleucine, leucine, lysine, methionine, phenylalanine, threonine, tryptophan, and valine— that are not synthesized by our body and, therefore, dietarily essential nutrients. These are commonly called essential amino acids. Three amino acids are particularly vital—leucine, lysine, and methionine. If you get adequate amounts of these, you typically get the other aminos in adequate amounts. Therefore, focusing on protein sources that meet critical thresholds of leucine (~2.5g) with a supporting cast of essential amino acids is important. Choosing whole food options consistent with your preferences, health circumstances, and ethically driven dietary decisions will achieve optimal health.

People following a plant-based plan can still achieve adequate protein, with the understanding that plant protein may be only half as bioavailable. The protein contained in plants is for the plant and is often attached to fiber, so it is not easily digestible for us to extract all the protein. Researching and understanding this is important as more protein sources are made available. Science continues to point toward the capacity of individuals to personalize their approach to maintain flexibility. One final thought on plant protein sources, many of the manufactured sources come with many added ingredients, so choose wisely.

Typically, people will follow the Recommended Dietary Allowance (RDA) for protein. My understanding is that this is based on sedentary behavior and young adults. If you are moving and exercising, your body will want more protein. Research shows the healthiest protein intake is approximately 0.7 – 0.8 grams per pound of lean body mass. This is a standard minimum, especially if you do not know your genetics. Based on your energy expenditure and if you know your genetics, you may require more like 1 to 3 grams per pound of body weight. Using an app to measure how much protein you consume can be beneficial.

Once you have accounted for the grams of protein required for your body, you can look at fats and carbohydrates. Research shows that diets high in fat and low in carbohydrates and vice versa are equally effective; it's a matter of preference. However, there is some genetic nuance to this as an individual. Unfortunately, many people have learned to become scared of carbohydrates and try to eliminate them from their diet at every moment. Carbohydrates are the body's primary energy source, which helps fuel your brain, kidneys, heart muscles, and the entire central nervous system. Annette Frain, RD, program director with the Weight Management Center at Wake Forest Baptist Health, said, "Fruits and vegetables are good for us; they're high in antioxidants and full of vitamins and minerals. If you eliminate those, you aren't getting those nutrients over time."

There are three main carbohydrate types: starches, fiber, and sugar. Starches are typically called complex carbohydrates, while sugars are considered simple carbohydrates. The difference between complex and simple is that complex typically contains more nutrients than simple carbs, which is why people aiming to improve their body composition are more apt to pick complex carbohydrates and limit simple carbohydrates and sugars. It may also be hard to get enough beneficial fiber while cutting back severely on carbohydrates. That can lead to digestion problems (ranging from constipation to diarrhea), bloating, increased body fat, and even elevated cholesterol and blood pressure. Fiber is generally good for aiding and maintaining healthy body composition.

There are specific genes for carbohydrates. A high-impact SNP is PLIN1. In studies, when an individual had a particular variety of this gene marker and they eat more than 150 - 250 grams of carbohydrates a day, their waist size was smaller. They tend to burn fat with an increased intake of carbohydrates. This is especially relevant to ideal body composition. At the same time, people with an alternate gene composition do better with less than 144 grams per day carbohydrate intake. In addition, some individuals show a propensity towards improved body fat loss and cholesterol management with higher levels of fiber in their nutrition plan. So interesting! Do you know your gene expression? Should you be eating more or fewer carbohydrates? I happen to have the PLIN1 expression recommending more carbohydrates. I tend to keep my ingestion to the lower end of the high scale, around 150 grams, to help with metabolic and dietary flexibility. You may want to experiment to see what works for you or have your genetics read for a better understanding.

Let me also explain carbohydrates as this is a wide variety of food. I think of these primarily as vegetables, tubers, and fruits. This is mainly where I am getting my 150g of carbohydrates. "Processed" simple carbohydrates are bread, pasta, baked goods, crackers, chips, etc. I suggest you reduce these at the beginning of your journey but don't eliminate them for the sake of your sanity. It is hard enough to change your eating habits and think healthy most of the time, but be smart about your food choices. A healthy balance is the key. If you have little self-control with certain items (you know what they are), then keep those items out of the house. I may go to a local store for a cupcake, an ice cream, etc., enjoy it and then go home, where I am not tempted to eat more. To foil the "sugar tooth" gene.

Now that you have the protein and carbohydrates accounted for, you make up the rest of your energy consumption with dietary fat. There are some genetics to consider here, especially regarding saturated and polyunsaturated fats. As we age, a genetic link exists between saturated fats and cognitive decline. Depending on your specific genetics, you will want to keep saturated fat as low as 5% of your daily energy intake. For others, they can be a little more flexible with as high as 10% of daily energy from saturated fat for the best possible health outcomes.

Polyunsaturated fat also has a genetic component. Some people must reduce items with high Omega-6, such as nuts. This can take some work to figure out, but many of my clients can make the changes quite easily by limiting certain items in their food choices and then using more olive oil or avocado oil, which are monounsaturated fats. Remember, many high-calorie dense items we crave, such as cakes and cookies, are often high in saturated or polyunsaturated fats. Consuming them may be working against our specific genetics and turning on genetic propensities for neurological decline as we age.

Does this seem overwhelming? Understanding it takes time, as does developing an eating pattern that works for you. Years later, I am still working on it. I prioritize my increased need for protein first, which, let me tell you, is a little bit of work. Then I factor in my increased carbohydrate need and drizzle it all with olive oil. You don't have to understand it all at once, and certainly don't have to be perfect. Just start moving in the direction of being healthier. To be cliche, this is a marathon, not a sprint (there is genetics for that, too). You will live in your body for the rest of your life. Take one day at a time and become comfortable with the changes. It is about doing your best to live a life of ability and health for as long as possible. I am sure there a many trips, family reunions, and games with grandchildren you want to participate in in the future. How do you want to spend the last couple of decades of your life? That decision starts with the choices you make now.

Does it matter when we eat?

One of the things you can consider is the time you eat. This topic has also become a source of debate. You may have heard about intermittent fasting and the health benefits it can give you. I have used some of these techniques personally, not so much for body composition but for the other health benefits it purportedly provides. As you may have guessed, I am not a big fan of "diets" unless the "diet" has your name on it. We are all unique individuals with unique biology.

Mark Mattson, Ph.D., a Johns Hopkins neuroscientist, has studied intermittent fasting for 25 years. He says we have evolved to go for periods without food. In prehistoric times, back to those cavemen again, before humans learned to farm, they were hunters and gatherers who evolved to survive for long periods without eating. Hunting game and gathering nuts and berries took time and energy.

Dr. Mattson says that after hours without food, the body exhausts its sugar stores and starts burning fat. He refers to this as *metabolic switching*. Research has shown that intermittent fasting can do more than burn fat. Mattson explains, "When changes occur with this metabolic switch, it affects the body and brain." One of Mattson's studies published in the *New England Journal of Medicine* revealed a range of health benefits associated with these fasting periods. These include a longer life, a leaner body, and a sharper mind. "Many things happen during intermittent fasting that can protect organs against chronic diseases like type 2 diabetes, heart disease, age-related neurodegenerative disorders, even inflammatory bowel disease, and many cancers," he says. However, it is not for everyone. Some people should avoid fasting and consult a healthcare provider on this one.

Another method of timed eating based on your circadian rhythm and biology is Time Restricted Eating (eating only during waking hours with an eating window of 8-12 hours). Based on the research conducted by Dr. Satchidananda Panda of the Salk Institute. He has studied time-restricting eating based on circadian biology, a natural, internal process that regulates the sleep-wake cycle and repeats roughly every 24 hours. His studies showed that eating only during daylight hours caused overall improved body composition and decreased body fat even without a calorie deficit (That's awesome news for those who can tune in more with our bodies). He found that time-restricted eating mediates the immune system, suggesting that genes and molecules involved in the circadian clock could affect inflammation-related conditions.

Circadian rhythms are everywhere in every cell ... We found that time-restricted eating synchronized the circadian rhythms to have two major waves—one during fasting and another just after eating. We suspect this allows the body to coordinate different processes.

Dr. Satchidananda Panda

According to research, your liver metabolizes glucose during daylight hours, is strongest in the morning, and weakens as evening approaches. Therefore, cells are much more sensitive to insulin during the day and less so at night. This is because melatonin will inhibit the release of insulin from your pancreas. Melatonin is not only a brain hormone but is made in your gut from Serotonin. Gut melatonin goes directly to the pancreas and modulates the secretion of insulin. Melatonin is secreted in most people three hours before they go to bed. Genetically for most people, this would be a bedtime of 10 PM. That's why it's important to eat lighter meals in the evening and stop eating at least three hours before bedtime.

Research on Time Restricting Eating showed that it not only affected how much energy stores participants burned but also increased fat burning for several hours at night and improved metabolic flexibility, which is the ability to switch between burning carbohydrates and burning fats. This is amazing!

Dr. Panda says, do not skip breakfast. You do not want to start your clock at noon. This goes against your circadian biology. Your circadian clock starts when you consume anything other than water. So even black coffee will activate the clock genes in your liver. Make lunch your biggest meal of the day. Stop eating after a maximum of 12 hours. For example, if your first cup of coffee or tea is at 6 AM, stop eating and drinking at 6 PM (except for water). This plan is a 12-hour/12-hour split and would lean towards eating larger meals earlier in the day.

Time Restricted Eating works with your biological clock and when metabolism is more active. When wearing a CGM, you can see some of this at work. Remember that Indian meal I ate? I learned a lot from that meal. That meal was in the evening, and my glucose stayed elevated longer than normal. I had a better response to my food in the morning. As the day progressed, I could see the difference in the glucose elevations and duration. Typically, I would start eating around 9 AM and stop around 7 PM, three hours before bed, with a nine to ten-hour eating window.

My research shows that time-restricted eating works better with your body's natural rhythm and life. Again, you can experiment based on your unique epigenome. Also, an added benefit of reducing hours of eating is that it helps you stay within a balanced energy burn by reducing the opportunity to eat more food if that is your goal.

Some final words on eating. I have been thinking about this a lot lately. It is easy to say we need to eat healthier while breaking bad habits. But it can be difficult. Why is that? Well, it comes down to our brains and our reward centers. We are programmed to find the most calorie-dense, rewarding food we can! Now you're asking why our brain would do that to us. Well, many moons ago (before grocery supercenters), we foraged and hunted for our food, back to those cavemen again. Our brain wanted us to find the most calorie-dense food from our environment to maximize our survival. Of course, the calories consumed were offset by the miles they walked looking for and hunting that food. Fast forward a few millennia, now we have an abundance of food everywhere, and the packaging screams to our brain, "calorie dense and appealing!" In some ways, it is hard to override those primal instincts when food designers are packaging them by using our chemistry against us. Then, we are sitting at our office job and moving less than we ever have (double whammy).

Suppose you're currently eating the Standard American Diet (SAD), a diet that has a combination of high carbohydrates, processed sugar, and saturated fats diet that can lead to an array of chronic diseases. In that case, it can be challenging to shift to a consistently healthy way of eating.

How can we combat shiny packaging and marketing? How do we not eat "SAD?"

1. Get rid of all the temptations lurking in your pantry and fridge. Do a "kitchen reset" and ditch the junk food, processed items, and sugary treats. The bottom line is don't keep tempting foods in the house. These shifts will benefit your whole family! Foil that "sugar tooth" gene if you have it.

2. Plan ahead. Plan your meals for the week or at least the next few days, so there's no guessing or "in the moment" lapses because you don't know what to eat. This will make your trip to the grocery store easy and ensure you have what you need on hand for healthy meals throughout the week. Here is a trick, shop around the outside aisles. The packaged, chemical-filled foods are in the middle section of the store!

3. Prep in advance. Prep your meats and veggies over the weekend to ease cooking in the evenings. For breakfast, make your protein-packed smoothies and freeze them or make an egg frittata or boil eggs for a quick, protein-rich start to your day. Aim for 30g of protein for each meal or base this on your protein genetics and LBM goals. Planning and prepping ahead of time will save you time and money and help you stay on track.

4. Only buy and cook with healthy oils. Use avocado for high heat and extra virgin olive oil for everything else!

5. Slow down and chew. Your brain knows when you are full because the stomach signals to a part of the brain called the hypothalamus. The hypothalamus is 12 minutes "behind" the stomach. So, there is a 12-minute delay between being full and your body knowing you are full. Be mindful and eat more slowly. You'll feel full before your plate is clean, and you will eat less.

6. Enjoy your food, and don't look at food as "good" or "bad." Our genetics and epigenetic expression can reveal which food choices align with your personal goals.

Start with baby steps. Even if you only replace one soda daily with a glass of water, you are improving your energy and nutrient intake. If you eat fast food every day for lunch, try replacing that half of the time with better quality, non-processed foods you bring from home.

When you start by taking small steps, you aren't cutting out all the foods you love all at once, and you'll be more likely to stick with your small changes. And keep in mind that these changes will positively impact your health and the health of your entire family for generations to come. The more health-promoting choices you make, the more you will crowd out the other options.

Creating new habits takes time, focus, consistency, and preparation. Remember, 1% better every day adds up at the end of the year. This can sometimes be challenging, especially if you have the "sugar tooth" gene or the "cookie jar" gene. I have both, so I understand.

As I chose to be healthy (not just go on another diet), I soon realized that eating for my genetics is a life choice. Although we may occasionally indulge in those SAD foods, eating high-quality, healthy whole foods is paramount.

Because this is not another "diet," you will occasionally indulge in what you are really craving. Have a small portion of what you crave but eat it slowly and mindfully. Enjoy your choice and then get back on track without any feelings of guilt that may sabotage your overall goal of health. If you have a craving that won't budge, tune in because your body often tells you it needs some specific micronutrients. And, while your brain needs glucose, it doesn't mean it wants refined sugar!

Okay, that is enough about eating for now. So much more could be said, but this is a good base for you to get started. Yet, I would be remiss if I didn't highlight the importance of energy balance.

The energy balance theory states that body fat gain or loss is determined by the balance between the energy consumed (calories from food and drink) and the energy expended (calories burned through physical activity and metabolism). When energy intake (calories consumed) is greater than energy expenditure (calories burned), typically, a gain in body fat will occur. When energy expenditure exceeds energy intake, body fat loss will occur.

This theory is based on the first law of thermodynamics, which states, *energy cannot be created or destroyed, only transferred or converted from one form to another.* In the context of the human body, this means that the energy consumed through food must either be used for energy expenditure or stored as body fat. This theory is commonly used to maximize optimal body composition strategies. Yet as I've said, it is important to note that other factors, such as genetics, hormones, and muscle mass, also play a role. We are complex humans in a complex world. Regarding hormones, you will get a greater understanding of this in Kym and April's chapter, especially as it relates to menopause.

As we eat energy, we must also burn energy. As I said before, most energy you burn happens naturally as you go about your day. But we also burn energy with increased movement. This was a challenging part for me. But I have learned to appreciate the feeling that movement gives me. I found things I enjoy, walking in nature, hiking mountains to see a view, and kayaking a lake or river to get a new perspective. This helps me burn energy, reduce stress, and maintain fat loss.

As I progressed in my health journey, I added some things to my routine as I became aware of the impact on my genetics. First, I added resistance training or weightlifting. This is immensely important for maintaining and building muscle mass and preserving health span and lifespan. It also helps you burn more energy stores (body fat). The second is cardiovascular exercise. For some people, Zone 2, and for others, higher, like Zone 4 and 5 (consult your physician). The cardiovascular system needs this, as does your heart. Remember, your heart is a muscle, too. I once watched an autopsy that I will never forget. Because of the lack of cardiovascular exercise, the person's heart was thin, like a paper bag. I thought, I never want my heart to be like that, so I get on the bike at least once a week and ride away, building that heart muscle.

The bottom line, if you are moving and not sedentary, you are halfway there. If you barely move because of a sedentary lifestyle or job, DO NOT underestimate the power of walking! The most significant needle move for you could be just walking more and finding things you love and enjoy increasing your desire to move more. I already know what some of you are thinking. I can't exercise; I can't weight train. But you can start to move by increasing your steps, even if by just 100 steps daily. You can purchase two-pound dumbbells and begin with one rep a day, then two reps, and so on, increasing your strength.

I vividly remember talking about exercise to a man who looked fit and healthy. He said to me, "walking isn't exercise!" How wrong he was. When I was nearly 300 lbs., walking was the only comfortable physical activity I could do. It helped me burn energy and begin to shift body fat. As the body fat shifted, I tried other activities, leading to new adventures. Physical activity is not all or nothing. It is about starting the journey and taking the first step. It is how I started. Did I wake up one day, go to the gym, and lift weights? No! Remember, I was the girl who disliked Phys Ed in school. I worked up to it, physically and mentally. You should, too. If you are younger, you may adapt to the changes of adding movement more easily. If you are older and this is a new lifestyle, you may have to see how things go. I love swimming as another great way to move, and there is little to no impact on the joints. If your pool is shallow enough, I have found running in the pool to be an excellent activity!

Remember, we can get crazy over specific nutrients, macros, and eating times. Still, *research shows that nothing is more powerful than an increased physical activity level for the best epigenetic modification.* I had to hit myself over the head with that a few times!

We have covered breathing and eating (epigenetics and energy). Now it is time to get some ZZs. Sleep is something most people are in desperate need of.

Eat (epigenetic and energy) Keynotes:

- Understand what eating style works for your unique genetic blueprint. Genetic testing can provide a wealth of information.
- Start with moving and walking. Strive for one percent improvement each day.
- Physical activity is the most significant lever to pull for healthy outcomes and maintaining energy balance.

Sleep

Sleep is like the golden chain that binds our health and body together.

Thomas Dekker

Sleep is foundational! Sleep quality is one of the most significant longevity and health span predictors.

Sleep is rarely mentioned as a pillar of health, yet a considerable body of evidence suggests that good quality sleep may be just as important to one's overall health as other lifestyle factors such as diet and exercise.

Judith Carroll,
an assistant professor of psychiatry at UCLA,

It makes sense when you think about it—we spend a third of our lives sleeping. Yet, we may only exercise an hour or so a day. Therefore, logically it makes sense that we need to get our sleep. It is one of the most important factors to consider when evaluating chronic disease. If a client has chronic health or metabolic issues, I look for the reason(s), sometimes connected to their sleep patterns. Sleep is crucial for metabolic functioning, hormone balance, and even immunity. Also, it makes us more zestful and more present when we get our sleep.

Sleep deprivation numbs you to your environment. It decreases your whole purpose for being here.

Dr. Kirk Parsley, Sleep Expert

It's all about squeezing as much enjoyment out of your life as possible. We were created to have the best life ever, full of discovery and adventure.

We have clock genes which are a group of genes regulating the circadian rhythm. Clock genes are the body's internal 24-hour "clock" that controls various physiological processes, including sleep. These genes help to control the timing of when we feel awake and when we feel sleepy. Each person has a genetic predisposition to how they sleep; some people genetically sleep less or more. Some wake up more frequently throughout the night, while others are less restless. However, with epigenetics, we can work on adjusting these genetic tendencies if they don't fit or if they disrupt our lifestyle. Disruptions in clock genes can lead to sleep disorders such as insomnia or jet lag. The disruption of these genes may also play a role in the development of certain mood disorders.

Generally, there are some recommendations for minimum time to sleep, as cited on SleepFoundation.org.

Recommended Hours of Sleep

Newborn	0-3 months old	14-17 hours
Infant	4-11 months old	12-15 hours
Toddler	1-2 years old	11-14 hours
Preschool	3-5 years old	10-13 hours
School-age	6-13 years old	9-11 hours
Teen	14-17 years old	8-10 hours
Young Adults	18-25 years old	7-9 hours
Adult	26-64 years old	7-9 hours
Older Adults	65+ years old	7-8 hours

Adults need to get a minimum of seven hours of sleep daily. Fundamentally, you are shortening your life and health span with less sleep. You must set yourself up to maximize your sleep. This is where working with your genetics and a sleep coach can pay off.

Quality sleep is the foundation that all other pillars of health sit on, and it supplies the energy needed to exercise, eat the right foods, and control your weight. The ideal place to begin improving your health and sleep hygiene is to reinforce your circadian rhythm with good light exposure before 10 AM. Your sleep schedule should be as consistent as possible every night, even on the weekends. Creating a nighttime ritual can signal your brain that it's time to slow down, rest, and start the melatonin release (the sleep hormone). Consider not eating at least three hours before bedtime, as discussed within circadian eating. Exercise earlier in the day or at least three to four hours before bedtime; otherwise, your body revs up, leading to sleep disruption or delayed sleep onset. Keeping all electronics out of the bedroom will also assist with sleep. The artificial light created by devices and TVs can delay melatonin release. Each hormone has a cycle and interacts with another hormonal counterpart. If melatonin is disrupted, this can create a cascade of effects leading to dysregulation in other cycles of hormones, such as Serotonin, the waking hormone.

Then look at the environment, such as keeping the bedroom cool. Your partner can always have a blanket if they need it. Also, keeping the room dark reduces stimuli that can cause excess waking. Often if clients like tea and it doesn't cause an excessive need to urinate at night, I suggest drinking some before bed. I have found several helpful and often mixed two bags. Combinations like chamomile and passionflower can help signal to your body that it's time to calm down. I have also found Valerian, catnip, and ashwagandha to be helpful.

If we look at our genetics, while everyone is different, there is a good chance your ideal sleep schedule would be from 10 PM to 6 AM. However, there are genetics for evening propensity. I was sure I had that! I had always preferred staying up late and waking up late. I was surprised to see that my genetics leaned more toward the morning. How could this be? As I reviewed my SNPs (Single-nucleotide polymorphism) within my genetic report's sleep panel, a couple had an evening preference. I must be expressing those, I thought. But then I found an SNP that increased morning fatigue. Yes, I definitely had that and the t-shirt that said, "I don't do mornings." Sometimes the clock of life isn't set up for my preferred sleeping schedule of 11 PM to 7 AM. Therefore, I strive for seven hours of good-quality sleep, even if it means going to bed earlier.

The human body has two phases of sleep cycles, (1) rapid eye movement (REM) and (2) non-rapid eye movement (NREM) sleep, which is further divided into three stages, wake, light sleep, and deep sleep. Each phase and stage of sleep includes variations in muscle tone, brain wave patterns, and eye movements. REM sleep is the stage of sleep during which the most vivid dreams occur, and it is also the stage where the brain processes and consolidates memories.

There are a few topics connected to sleep that I would like to highlight that can optimize and help you thrive in life. Each of these topics could be a whole chapter in itself. If any of these pertain to you, there are many resources to learn more about your specific sleep genetics.

First, let's talk about sleep as it's related to maintaining healthy body composition. When most people think of getting healthy, they first think about losing a few pounds. But, as you will see in Laurie Kaplan's Chapter, a "diet" isn't necessarily the first place to start. Chronic sleep deprivation, or losing 30-60 minutes of sleep daily, increases your risk of gaining excess body fat. Your resting metabolic rate (RMR) is how much energy your body burns at rest. RMR is affected by age, weight, height, sex, and muscle mass. Research indicates that sleep deprivation may lower your RMR. This reduces the energy you can consume from food in a day and can put you at risk of consuming excess energy, which would then be committed to storage (body fat). Therefore, your "recommended calories for the day" will not be entirely accurate for those who perhaps have a wearable and struggle with sleep. In other words, if you're not sleeping well, it could mean that your RMR could be lower and not accurately reflected on your device. This can mislead you into thinking you have more leeway in your energy consumption and potential over-consuming of food. RMR is also affected if you have yo-yo dieted for years, as I had. Furthermore, research indicates that you can lose muscle mass if you're sleep-deprived. Of course, as discussed in the Eat section, we know that more muscle mass burns more energy, and maintaining muscle mass is vital to long-term health.

Are you starting to see a pattern? Here is human complexity at its best or worst, depending on your worldview. All of these factors intertwine and indicate that if you are not getting your sleep, you're at a disadvantage, especially if you are trying to maintain healthy body composition. The truth is you've got to get a minimum of seven hours of sleep; otherwise, it will be harder to maintain or take excess body fat off. You won't necessarily lose body fat if you get more sleep, but too little sleep hampers your metabolism and contributes to gaining body fat.

To add further complexity, your brain's ability to function on less sleep contributes directly to your physical and mental health outcomes. Adequate quality sleep is essential for cognitive performance, especially memory consolidation. People who experience sleep loss usually experience a decline in cognitive performance and changes in mood. Sleep deprivation seems to activate the sympathetic nervous system (the system activating the "flight or fight" response), which can increase blood pressure and cortisol secretion. Your body's immune response may be impaired, and metabolic changes such as insulin resistance could occur.

A lack of sleep disrupts brain functioning, primarily in the frontal lobes, affecting our decision-making ability and impulse control, leaving us reaching for comfort foods and leading to mindless snacking. Lack of sleep often increases our desire for those lovely calorie-dense carbohydrates and weakens our resolve not to eat those snacks. You can see this leads to the perfect storm of overconsumption. Just look at the morning line at the various sugar and caffeine vendors; plenty of people lined up for their fix. I would guess that if we surveyed them, most do not sleep well. Watch this in yourself the next time you do not sleep well. Are you craving the sugary "pick-me-up" more than usual? Let's discuss why.

A hormone key to hunger signals is ghrelin, which is released in the stomach and signals the brain that we are hungry. Another important hormone is leptin, which is released from our adipose tissue (body fat). Leptin is a signal to your brain that lets your brain know how much body fat you are carrying. As discussed in the eat section, normally, leptin suppresses our hunger signals and tells us we do not need to eat as we have enough fuel. However, if we don't get enough sleep, our adipose tissue starts to act like there is insufficient fuel for the body. It then sends the signal (leptin) to the brain to eat more. Even after one night of sleep deprivation, these hormones get disrupted, and they can make you crave more and increase your appetite for primary carbohydrates. In a study of over 1,000 people, researchers found that if participants had a shorter sleep duration, ghrelin was increased by 15%, and leptin was lowered by 15%. It also indicated that those with shorter sleep cycles tended to carry more body fat. To maintain healthy body composition, it is essential to maintain a minimum of seven hours of sleep.

As a teenager, I slept great, I used to say a train could come through my bedroom, and I wouldn't notice. Then as I got older and more stressed, sleep became more difficult. I struggled with early morning awakening and frequent sleep disruption. Once I was thrown into menopause, sleep became an even more significant issue. In the early stage of menopause, I would wake drenched in sweat, and then the hot flushes would begin. At times I thought I would spontaneously combust. Hot flushes became so regular that they were like an alarm, waking me up around the same time each day. Of course, my body set this alarm at the most inconvenient times, like 5 AM. I became desperate to have a good night's sleep.

Once I understood my sleep, nutrition, supplementation, hormone genetics, circadian eating, and other principles, I could hone my sleep.

I used supplements to reduce the hot flashes, worked on my sleep hygiene, stopped eating so close to bedtime, and changed the types of food I was eating.

Dr. Kirk Parsley recommends that deep sleep and the REM cycle be 25-30% of your total sleep. In other words, if we sleep a minimum of seven hours, we should get at least one hour and 45 minutes of deep sleep, the same for the REM cycle. I mentioned that more sleep issues began after menopause onset. While I don't hit these percentages some nights, I am much closer and more consistent than I was during the early years of menopause. Typically, I can pinpoint why these stages were lower than they should be or more disturbed, such as something I ate or drank, the day's stress level, working on the computer, or viewing electronics too late.

I also worked with my sleep genetics to reduce the number of micro awakenings I have during sleep, which I have genetic propensities towards. These are short periods of arousal, typically for a few seconds, in which you wake up, change your position, and go back to sleep. If disturbed, micro awakenings can last up to a few minutes or occur several times. In the early days, these could add up to over an hour or more of lost sleep for me. Most nights, I can increase my deep restorative sleep and REM cycle while reducing micro awakenings. I also have the genetics for a shorter sleep cycle overall, which has taken more work to change. Yet, if I follow my sleep schedule, I can achieve at least seven hours of sleep.

Unfortunately, sleep issues are common with menopause. Looking at the statistics in premenopausal women, 47%, almost half of the women, have problems with sleep. In postmenopausal women, it increases to 60%. I was not alone in my struggles; sleep is a big issue for women around the age of menopause. The two most common sleep disruptions reported are hot flashes or night sweats and insomnia.

When those night sweats hit, it is because of changes in estrogen. Estrogen plays a significant role in the metabolism of serotonin, which plays a key role in such body functions as mood, sleep, digestion, nausea, wound healing, bone health, blood clotting, and sexual desire. When you experience a hot flash, it can increase the adrenaline in your body, and it can make it even harder to go back to sleep. Of course, every awakening from a hot flash affects the overall sleep quality. It becomes uncomfortable, causing real fatigue, and eventually, if it becomes chronic can cause insomnia.

One of the first things I do with clients is to avoid triggers that can induce an increased frequency of hot flashes. For example, consuming alcohol, nicotine, or caffeine can increase the frequency of hot flashes. If they love their coffee or drink soda, it is recommended that they minimize consumption as much as possible. Perhaps one cup of coffee early in the day. Limiting or avoiding caffeine is best if they are having trouble sleeping, especially in the evening.

Caffeine is a stimulant that can affect sleep by delaying the time it takes to fall asleep and reducing the overall time spent in deep sleep. It works by blocking the action of adenosine, a neurotransmitter that builds up in the brain over the course of the day and promotes feelings of sleepiness. The half-life of caffeine is around four to six hours. This means that if you consume caffeine at 6 PM, half of it will still be in your system by 10 PM, which can cause sleep disturbances and make it harder to fall asleep. It is recommended to avoid consuming caffeine close to bedtime.

I might also recommend avoiding spicy food, especially close to bedtime. Eating spicy food can affect sleep in a few different ways. Firstly, spicy food can cause acid reflux, making it uncomfortable to lie down and sleep. The acid from the stomach can flow back into the esophagus, causing a burning sensation in the chest or throat. This can make it difficult to fall asleep, especially if our eating plan is not aligned with our genetics. Secondly, spicy food can also increase body temperature, causing sweating and discomfort, making it difficult to fall or stay asleep.

This is compounded by menopause with the advent of night sweats and hot flashes. Lastly, spicy foods can cause an increase in heart rate and blood pressure, which can make it harder to fall asleep. Keep in mind that the effects of spicy food on sleep can vary from person to person, and it's always best to pay attention to how your body reacts to it.

Drinking alcohol can disrupt sleep in several ways. Firstly, alcohol is a sedative that can help you fall asleep more quickly but also interferes with the normal sleep cycle. This can lead to a lighter, less restful sleep and can cause you to wake up more frequently throughout the night. Additionally, as the body metabolizes the alcohol, the sedative effects wear off, which can cause a person to wake up in the middle of the night and have trouble falling back asleep. This is known as "rebound insomnia." Alcohol can also cause snoring and sleep apnea, further disrupting sleep.

Additionally, drinking alcohol can affect the amount and quality of REM sleep. Alcohol consumption can increase the time spent in the initial stages of non-REM sleep, prolonging or suppressing REM sleep. This means that a person may not experience as much REM sleep or may experience it later in the night after the body has metabolized the alcohol. This may result in less restful sleep and lead to memory and learning problems. Moreover, alcohol can also cause vivid and disturbing dreams, further disrupting REM sleep and leading to poor overall sleep quality.

How does sleep affect our epigenetic clocks? Judith Carroll, from UCLA, states, "Not getting restorative sleep may do more than affect our functioning the next day. It might also influence the rate at which our epigenetic clock ticks," Epigenetic clocks are methods of measuring biological age based on the patterns of methyl modifications to our DNA. Again, affecting how genes are expressed. Epigenetic clocks have been shown to be associated with various age-related diseases and conditions, including inflammation. Epigenetic clocks have been used to study the relationship between inflammation and aging. For example, research has suggested that epigenetic age acceleration, which measures how much a person's epigenetic profile deviates from their chronological age, is associated with inflammation and a higher risk of age-related diseases.

Carroll said. "In the women we studied, those reporting symptoms such as restless sleep, repeatedly waking at night, having difficulty falling asleep, and waking too early in the morning tended to be older biologically than women of similar chronological age who reported no symptoms." This information was found using a genetic "biological clock" developed by Steve Horvath, a professor of human genetics and biostatistics, which has become a widely used method for tracking the epigenetic shift in the genome. "We can't conclude definitively from our study that the insomnia symptoms lead to the increased epigenetic age, but these are powerful findings," Carroll said. "Our previous research found that one night of partial sleep deprivation promoted cell damage that can increase susceptibility to biological aging, suggesting a causal connection." These findings are not all bad. Knowing that epigenetics can be modified means we can change behaviors to reduce overall risk by improving sleep quality each night.

One of the most interesting discoveries in the past decade is that the brain has a "waste management system." During the day, your brain cells need to eat (to absorb, primarily, glucose and oxygen). Like any other "digestion" process, waste products must be disposed of. The brain's waste management system, called the glymphatic system, takes out the trash for the brain at night. It is a "pseudo-lymphatic" perivascular network distributed throughout the brain, responsible for cleaning and revitalizing the brain.

Our lifestyle choices, such as sleep position, alcohol intake, exercise, omega-3 consumption, intermittent fasting, and chronic stress, modulate glymphatic clearance. If this system starts to under-function due in part to our lifestyle choices, including chronic sleep deprivation, there is growing evidence that it plays a role in neurodegeneration and may even play a role in other brain disorders, including Alzheimer's disease. Research shows that sleep deprivation contributes to the accumulation of metabolic wastes and toxins in the brain, leading to dramatic glymphatic and meningeal lymphatic impairment.

In addition, a disturbed sleep-wake cycle increases the risk of cognitive decline and diseases affecting the small arteries, arterioles, venules, and brain capillaries. This becomes even more important with the genetic variation of APOE (apolipoprotein E). In the show, *Limitless*, with actor Chris Hemsworth, they discussed how his genetic variant of APOE gave him the propensity to be "eight to ten times more likely" to develop Alzheimer's disease eventually. This naturally caused him to reflect on his health and mortality. He immediately stepped back from working a relentless schedule and began practicing lifestyle changes and reducing stress, including time with friends and family. If we want to increase our body and mind vitality, managing our sleep and stopping burning the candle at both ends is important, as it is so easy to do in this busy life we have created.

I would be remiss if I did not mention how stress and worry affect sleep. After all, I am sure you, like me, have had a sleepless night from worry. I am not referring to chronic anxiety; this is more of acute stress and worry that keeps you up at night. The challenge with stress and sleep is that they are bi-directional regarding how they interact. It's like the chicken and the egg - what came first. Stress and sleep impact each other. Reducing stress should become a priority if you are experiencing sleep issues. Look at the relationship that stress has on you and if it is causing your sleep disruption. Are you sitting up at night because of problems or worries? Or are you stressed at not sleeping regardless of what happened in the day? It may be hard to tease apart. Chronic sleep deprivation or interrupted sleep can lead to higher cortisol levels and ironically make you feel wired at the end of the day, unable to unwind naturally. If stressing is compounding sleep problems, it can put you at higher risk for metabolic dysfunction. If you have stress, and if this affects your day-to-day life, you should speak with your healthcare provider.

Hormesis is the phenomenon where low stress levels can benefit cellular processes and overall organism health. These low-level stresses can trigger changes in gene expression through alterations in DNA methylation and histone modification. These changes can result in improved stress tolerance and resistance to disease. However, now for that wooly mammoth! As I've suggested before, we have an ancient aspect of our brain that goes into "flight or fight" mode; it's an ancient survival mechanism. Though no T Rex or woolly mammoth is coming through the door (cave entrance?) anymore, the past echoes when we overthink bills, mortgages, car payments, or worry about the kids. It's all about the perceived threat. Over decades, our brains have generalized these worries and sense of danger. We must be aware of it and develop an escape route if it creates a sensation. We are right back to that low level of chronic stress again.

It's a great idea to evaluate what is causing anxiety and how that may translate into issues at night with sleep. The occasional anxiety is not a big concern, but chronic stress and how it relates to our sleep is a cause for concern. I often suggest journaling about their worries or stressful issues when working with clients. This simple act can help distance or remove the subject from their mind altogether. Explore the subject thoroughly until it is exhausted. Meditation is another great way to relieve anxiety before bedtime. Again, the apps work well. I highly recommend Sam Harris' *Waking Up* app. He talks a lot about imagining the thoughts unraveling in your mind. The healthy brain has an immense capacity for resilience, with epigenetic modification and direction toward better health. While there are no magic bullets (except the ones for smoothies), mindfulness-based stress reduction and meditation are valuable tools.

Other ways to release stress at night before going to sleep are essential oils. Lavender oil is one of the most researched oils to improve sleep and relieve stress. Lavender oil used in a carrier oil can be put right onto the skin, pillow, or diffuser. Massages, regular exercise, and stretching exercises can relieve stress and help sleep. Physical activity makes the body tired, increasing the need for more sleep to aid recovery. Sleep is where the recovery occurs.

When I work with clients, I often start by assessing their sleep. Review their genetics and sleep history to understand how it may impact them on a wider level. As you can see from the discussion above, the body's metabolic functioning and the brain's capacity to function normally depend on good-quality sleep. The amazing thing about epigenetics is that we can monitor the changes and evaluate therapies. We could even improve our biological age by changing our sleep quality and length.

To find helpful ways to improve your sleep, email me for a copy of my Sleep Enhancement Worksheet at office@DrRachelleSweet.com.

Sleep Keynotes:

- Establish a regular sleep schedule: Go to bed and wake up at the same time every day, even on weekends, to help regulate your body's internal clock.
- Create a sleep-conducive environment: Keep your bedroom dark, quiet, and cool. If necessary, use earplugs or a white noise machine to block out any noise. Use comfortable bedding, and make sure the mattress and pillows support your body.
- Wind down before bedtime: Give yourself time to relax before bed. Avoid screens, bright lights, and stimulating activities for at least an hour before bedtime. Avoid consuming caffeine, alcohol,

and heavy meals close to bedtime. Consider using a relaxation technique such as progressive muscle relaxation or deep breathing exercises to help you fall asleep.

- It's also important to maintain healthy habits throughout the day, such as regular exercise and early exposure to natural light before 10 AM, to improve the quality of your sleep.

Thrive

To fully thrive, we must eliminate the stressors and actively seek joyful, loving, fulfilling lives that stimulate growth processes.

Bruce H. Lipton

It may sound cliche to say, "don't just survive, thrive." Something you might see on a bumper sticker. To thrive is to progress toward or realize a goal despite or because of circumstances. All of us can get behind this. As beings interacting in this world, most of us are progressing towards something better, striving for something more. Certainly, if you have picked up this book and read it this far, you are already on the road with us.

In some sense, our epigenome also represents our ability to adapt, evolve, and thrive by expressing characteristics or phenotypes developed in response to our environmental stimuli. Your epigenome is at the heart of your ability to thrive. Ultimately, the environment presents various factors to us, and each has an individual influence on our epigenome. As each individual has a unique epigenetic and genetic profile, we can modulate the specific response to these environmental factors to improve the quality of our health and well-being and, ultimately, our ability to thrive.

Thriving, flourishing, or prospering can be applied to different aspects of life, such as physical health, mental well-being, and social and economic success. Within the field of epigenetics, there is a relatively new connection between thriving and epigenetics. As discussed, certain epigenetic changes can affect the expression of genes involved in inflammation and the stress response. Learning to turn these genes down can positively impact overall health and well-being.

The growing body of evidence about lifestyle factors leading to beneficial epigenetic changes to promote health and well-being grows daily as our perspectives change and expand. Because of our complex world, we are constantly impacted by environmental changes. This makes the connection between epigenetics and thriving complex but also personalized. For example, an epigenetic change beneficial for one individual may not have the same effect on another, just like what makes one person engage in one worthy activity may not make another engage in the same activity. Understanding how these environmental factors influence the individual epigenome and how the epigenome, in turn, modulates an individual's response to these factors is crucial in understanding how to promote health, enjoyment, and longevity.

This leap will involve changing our perspective multiple times in this journey we call life.

> *The world we have made as a result of the level of thinking we have done thus far creates problems that we cannot solve at the level of thinking at which we created them.*
>
> Albert Einstein

As we grow and expand, we are presented with obstacles. These obstacles are new opportunities. We must change our thoughts and perspectives to improve what is in front of us. Our mindset or outlook on life is the main factor behind whether we pursue or stick to our goals.

During the psychological state of thriving, people experience a sense of both vitality and learning. Thriving individuals are growing, developing, and energized rather than feeling stagnated or depleted. Research has found that people who experience a thriving state are healthier, more resilient, and more able to focus on their work. In addition, people who feel like they are thriving are buffered from distractions, stress, and negativity. This is ideal for reaching a sense of calm and positivity.

Is there such a thing as a happiness gene? When researchers looked at a twin study, the results showed that participants with a higher presence of the number of longer alleles of the 5-HTTLPR gene (a serotonin transporter gene) self-reported higher levels of life satisfaction. While the investigators did not define this gene as the "happiness" gene, it seemed to correlate 33% of subjective life satisfaction with genetic variation. Understanding that in some way 'happiness" is due to genetics gives you a remaining 67% to decide how happy you want to be. This choice allows room for the release of comparison and competition and leaves room for abundance instead. A person's base level of happiness appears to be governed by three major factors: a genetically determined "set point" for happiness, circumstantial factors, and activities and practices. Several researchers have argued that the ability to be happy and content with life is a central criterion of adaptation and positive mental health.

Within the concept of mindfulness, I have discovered that the idea of "happiness" is often fleeting; the idea is often temporary and tends to fade over time. The concept is that people tend to return to a baseline (genetic) level of "happiness" after experiencing positive events, such as receiving a promotion or getting married. This process is known as hedonic adaptation, which refers to the tendency for people to adapt to new circumstances and for the initial "happiness" derived from the event to fade over time. Rather, the focus is on living in the present moment and experiencing joy and contentment rather than constantly seeking happiness in some future event or dwelling on the past. It brings into focus what you have currently and allows you to appreciate it rather than focusing on what you lack or what you wish you had. It means finding joy and contentment in the little things in life, such as spending time with loved ones, enjoying nature, or pursuing hobbies.

Finding enjoyment and discovery in every moment is to seek an open awareness of the positive with a sense of exploration, curiosity, and purpose. Experiences of pleasure, engagement, meaning and purpose have been associated with positive epigenetic DNA methylation changes that promote improved mood, cognitive function, and physical health.

We understand that health is a fundamental aspect of well-being and closely relates to an individual's ability to thrive. Poor health can limit an individual's ability to engage in activities and experiences that promote well-being and can lead to negative outcomes such as increased stress and decreased life enjoyment. Wealth can also influence an individual's ability to thrive. Having sufficient financial resources can provide individuals with more opportunities to engage in activities and experiences which promote discovery and growth while reducing stress and anxiety associated with financial insecurity. Wealth can also provide access to resources, promoting physical and mental thriving.

Time is also an essential factor influencing an individual's ability to thrive. Having enough time to engage in activities and experiences, such as spending time with loved ones, pursuing hobbies, and engaging in self-care, can improve an individual's overall outlook and ability to thrive. Additionally, having a sense of control over one's time, such as setting boundaries and prioritizing activities, also affects thriving.

If any of these three areas (health, wealth, and time) are out of balance, the net effect is a decrease in enjoyment and thriving. For example, physical fitness, nutrition, financial security, and connections are crucial to overall health and well-being. However, spending excessive time in any of these areas will decrease the opportunity to thrive fully in another area. Finding a balance is a struggle—we know these stories too well. For example, spending all our time pursuing pleasure and enjoyment could leave us destitute if it doesn't provide income.

Conversely, pursuing income can lead to social destitution. It is the never-ending conundrum of balance. Ultimately, the key is to find a fulcrum between these three aspects of life. Each is interconnected, and each aspect, when aligned, can lead to overall thriving.

Yet, our life satisfaction has more to do with our experiences and memories with other humans than most people realize. How our brains perceive others and our ability to be resilient while being accurate in that perception is thought to be one of the pathways to provide a more positive way to view the world. Understanding self-acceptance while perceiving others with loving kindness leads to a flourishing life. The late Wayne Dyer had a saying that reminds me of this, "When we change how we look at the world, the world changes." Our worldview shifts when we shift our perspective and approach to every aspect of our life. The shift changes our neurochemistry, increasing neurotransmitters associated with positive emotion leading to a more fulfilling life.

While research has established that thriving is similar to resilience, prospering, or growing, it is different. To thrive, several areas must be considered: a positive outlook, a connection to something meaningful, whether it is spiritual, religious, or a relationship with a power greater than you (prayer, meditation, or spending time in nature), the desire to pursue a passion, challenging yourself, and motivation for growth.

Why not become curious about pursuing new ideas or skills, overcoming obstacles, seeing the benefits of a struggle, or seeing the lesson in adversity? Maybe connecting socially at a new level. It serves us to share feelings in the spirit of cooperation or connection.

A passionate mind keeps us purposed and relevant.
Yet, one of the most overlooked aspects of growing older
is that we are still growing.

Cynthia R. Green, Ph.D.

Being passionate about something can also bolster your thrive factor and longevity. Engaging in something you love to do has a slew of health benefits. For example, I love hiking and enjoy spending time in nature and seeing something beautiful. I love the "awe" factor. After the hike, I feel much better all around (even if I was not necessarily feeling down before the hike). The hike represents many things for me. There is the idea of communing in nature. I love to be surrounded by trees and be away from the hustle and bustle.

"Forest bathing" is a term used to describe spending time in nature to improve physical and mental health. Studies have suggested that spending time in a forest or other natural environment can have several benefits, including reducing stress levels, improving mood, and boosting the immune system.

There is also evidence that spending time in nature may have epigenetic benefits. Studies have found that spending time in nature may lead to changes in gene expression, such as reducing inflammation and promoting the growth of new blood vessels.

In nature, particularly in forests, there are high levels of negative ions in the air. Negative ions are atoms or molecules that have gained an extra electron, often found in high concentrations near waterfalls, on mountaintops, and in other natural environments.

It is believed that negative ions in the air can have many benefits. Some of the benefits include:

- Improved mood: Negative ions are thought to increase the levels of serotonin, a chemical in the brain associated with feelings of happiness and well-being.
- Reduced stress: Negative ions are believed to neutralize positive ions in the air, which are thought to contribute to feelings of stress and anxiety.
- Increased energy: Negative ions are thought to increase oxygen levels in the blood, which can lead to improved energy and greater alertness. This is especially true if you are breathing through your

nose. while also helping to reduce symptoms of asthma and allergies by neutralizing pollutants and allergens in the air.

As I hike, I believe these particular natural environments allow me to process my thoughts, a walking meditation. It assists me in alleviating stress and worry. I begin to feel a sense of calm and balance.

Hiking also is a way to overcome my struggle with exercise. Because of my positive perception and outlook on hiking, I am more apt to do it. It also challenges me personally and mentally if the hike itself is challenging. Terrain that requires using natural movement patterns, such as balancing, crawling, and climbing, strengthens the body and improves overall fitness. Not to mention problem-solving! Which can also create new neural networks in the brain. Many natural environments provide opportunities for overcoming obstacles like rocks, trees, and rivers that must be navigated. This type of movement can help to improve balance and coordination, increase flexibility and mobility, and build strength and endurance. Some of which you can't get in a gym.

Some of my most challenging hikes have provided a sense of accomplishment that helps build my self-confidence and self-esteem. Sometimes the hike provides an opportunity for self-reflection and introspection. Looking inside ourselves to identify what is happening within reduces stress and creates neurogenesis or more brain connections. Whenever you are doing something new, it stimulates the brain to make new neural pathways, which keeps you growing physically and mentally. Think about what you love to do that can provide these outlets in your life.

It is not enough to have a good mind. The main thing is to use it well.

Rene Descartes, French philosopher

Similarly, composing and writing this book was challenging mentally and emotionally. It made me grow in ways I had not thought of or expected.

It challenged me to overcome years of negative programming and beliefs about myself. I was told as a child that I would not go to college because my "style" of learning wouldn't get me there. However, I was not going to let that be my destiny. I battled against the "system" and carved an academic path based on my unique skill sets and abilities. I found hacks that work for me by using my brain power. The path was fraught with barriers and, at times, people who just felt I did not measure up. Yet, I persisted— evolving and adapting. Academically, I went as far as possible, culminating in a doctorate. I thought that was the top. I never thought I'd write a book, yet here I am. Because I believe it matters and is something that a power greater than me helped bring forth.

Finding your connection to a power great than yourself can also promote thriving. There are so many spiritual paths. I like to learn and assimilate knowledge from different views; I feel there is some merit to each doctrine's wisdom. Because of my interest in meditation, I have learned a little bit about Buddhism. The teachings focused on the path to enlightenment and the elimination of suffering.

In terms of thriving, it is thought that true happiness and well-being come from understanding the nature of reality and the mind and developing wisdom and compassion. The ultimate goal is to find a way to end suffering and find inner peace by developing wisdom and compassion within. It is believed that suffering can be overcome through mindfulness and meditation. As discussed in prior sections, studies have found that mindfulness practices can lead to changes in the methylation of genes related to stress and inflammation. Additionally, compassion towards others and loving kindness may also lead to changes in gene expression related to immune function and may improve the body's ability to fight disease. Choosing a loving-kindness meditation, often called Metta, is one strategy for boosting happiness. This strategy helps to increase oxytocin and reduce cortisol. Focusing on loved ones while relaxing self-focused thoughts increases overall well-being.

Compassion is empathizing with others and feeling concerned for their well-being. It is often considered a fundamental aspect of emotional well-being linked to positive mental health outcomes. Research suggests that being compassionate towards others can lead to increased feelings of satisfaction. Studies have found that more compassionate people tend to have greater emotional resilience and cope better with stress and adversity. They also tend to have higher levels of self-esteem and self-worth.

Practicing compassion can also have a positive effect on relationships and social connections. Social connection, such as volunteering, reduces stress, decreases depression, and boosts overall well-being. Volunteering keeps you cognitively engaged and neuronally flexible as you learn new things. This is vital as we age. When we connect with other like-minded people through social interactions, our brain releases oxytocin which gives us a calming, soothing feeling. People with higher oxytocin levels are more trusting, empathetic, and generous. When compassionate towards others, we are more likely to form stronger, more meaningful connections, leading to increased happiness and satisfaction.

Recent research on longevity suggests that certain aspects of emotional well-being, such as stress or positive emotions, can influence the epigenetic marks on our genes, leading to premature biological aging or the biological slowing of the aging clock. A study found that people who scored higher on a self-compassion scale had a different epigenetic profile in the gene associated with inflammation, suggesting that self-compassion may have an anti-inflammatory effect at the epigenetic level. Another study found that individuals with higher levels of self-compassion had different methylation patterns in genes related to stress and emotional regulation. It is important to note that self-compassion and self-love can also change how we perceive and react to stress, leading to improved coping mechanisms and enhancing epigenetic marks on our genes.

The importance of practicing self-compassion, which is being kind and understanding towards oneself when facing personal difficulties, rather than being self-critical or judgmental, is paramount. It is too easy to throw in the towel and spiral downward when facing adversity. Self-love and self-compassion are essential to emotional thriving and have been linked to several positive health outcomes. Studies have found that self-compassion is associated with greater psychological well-being, including increased happiness, decreased stress, and depression.

Visualize your best possible self. Then practice the behaviors that your best self would engage in. Over time and with consistent effort, you will become that best possible self. Stop and view how you speak to someone else before saying something critical or self-deprecating about yourself. Be kind to yourself. Choose language that is gentle and encouraging. Do not believe everything your mind says about you!

Gratitude is another emotion associated with positive outcomes, leading to improved relationships and better physical health. Research has found that individuals who practiced daily gratitude had changes in their gene methylation, which was associated with improved mood, less stress, and better sleep. A similar result was found by individuals who kept a gratitude journal, where changes in their gene methylation were associated with improved physical health and reduced inflammation.

It is easy to forget that our bodies have natural wisdom and intelligence. Have you considered that "incurable" may refer to something cured from within? Our bodies have an intrinsic knowledge of growing, healing, maintaining balance, restoring homeostasis, and regenerating. Western medicine emphasizes the chronic use of drugs to suppress the symptoms of illnesses. Don't get me wrong—there is a place for acute care and emergency surgery. There is a place for pharmacology. I would not be here without it. Yet, there is something flawed in this model. With our increasingly sedentary and hectic lifestyles, and with less emphasis or the wrong emphasis on nutrition and exercise, where are we headed? "Lifestyle-related" diseases, including obesity, diabetes, cancer, and heart disease, are rising despite numerous health initiatives, plans, and programs.

Genetics has been at the leading edge of research for several decades in understanding human disease. Now we are exploring epigenetics to understand health and well-being to promote longevity and thriving. Epigenetics establishes a scientific basis for how everything we do. Breathing, eating, sleeping, and thriving can shape us physically and mentally. When we take a holistic view of our health, not just a symptom-reductionist view, we can identify an epigenetic connection between intervention and gene expression, making it possible for humans to heal and grow beyond the limits that were previously perceived.

The field of personalized well-care aims to tailor treatments to the unique characteristics and needs of the individual client. Epigenetics is becoming increasingly important in this field. Epigenetics can be used to identify specific biomarkers that can be used to predict a person's risk of developing certain diseases or to determine how an individual will respond to a particular treatment to mitigate or turn down the influence of the risk. Via genetics, we can look at our individual response to sleep, nutrition, stress, hormones, the environment, athletics, and much more.

In the following chapters, you will read stories of those who sort other answers to their ailments because they felt traditional medicine did not entirely work for them. They found coaches or knowledge that helped them look at the complexity of the human system and go beyond it. Together with their coach and mentors, they worked to balance and heal their systems using epigenetic modulation. By understanding the fundamental genetic blueprint, they were born with, and how it has come to interact with the world, they made small changes that led to significant impacts on health and longevity in themselves and others. Some even liberate themselves from illness and pharmacology.

This awareness allowed the professionals, coaches, and clients in the following chapters to shape their habits, nutrition, environment, and other factors beyond the acquired genetic traits. As science and health care move into the future, this type of evaluation, genetic testing, complete blood analysis, and review of the psycho-social and physical environment is the future of well-care and essential to improving health and lifespan. By coming to know ourselves and what impacts our epigenetics on every level, we can learn to go beyond surviving to thriving.

Thrive Keynotes:

- Prioritize self-care: To thrive, it is essential to prioritize self-care and focus on maintaining good physical and mental health. This can include regular exercise, eating a healthy diet, getting enough sleep, and managing stress.
- Make time for activities and experiences that promote well-being: Engaging in activities and experiences that promote well-being, such as spending time with loved ones, pursuing hobbies, and traveling, can improve the overall quality of life.
- Seek balance: To thrive, it is important to seek balance in all areas of life, including health, wealth, and time. This can include finding a balance between work and personal time, maintaining a balance between spending and saving, and balancing physical and mental health.

Epigenetics of Consciousness

by Mickra Hamilton, AuD.

Consciousness rides the breath! For me
It is the foundational human systems focus.

Mickra Hamilton, AuD.

Could something beyond genetics shape awareness of ourselves and the world around us? How is this possible, exactly? Through the science of epigenetics, a concept that's often met with skepticism, we open the door to new perceptions and the potential for a bright and different future.

An interesting combination for sure, and you may wonder if this is a chapter exploring consciousness and epigenetics through the rather popular well-worn, new-age philosophy or solely through a rigid scientific lens. Fortunately, this chapter will explore it through the "both-and" lens of perception. This writing aims to provide a clear window of perception for the power we must create in a life of excellence. I invite you to discover an integral, complex systems approach to the epigenetics of consciousness.

The two intertwine to move in the most delicate embrace in this experience called life. We come into this mysterious world, emerging from infinite nothingness and filled with endless possibilities. From frequency to electromagnetics to matter, we move in an exquisite orchestration, oscillating between harmony and disharmony, unity and separation. All systems are interconnected and interrelated, with nothing standing alone or separate. As we journey through reality, immersed in illusions that deceive us into believing separateness exists when it doesn't, only love remains to be discovered within ourselves as well as connectedness among all things. To me, "consciousness" isn't just about thought processes or emotions. It is everything!

Biology, consciousness, and the universe are not fixed entities; epigenetics is a powerful driving force in their evolution. Epigenetics unlocks limitless possibilities. The topic is often misconstrued since it's gained so much popularity among "mystical thinkers" and marketers alike. Regardless, there is still an incredible amount of scientific research behind this concept that over 450,000 peer-reviewed articles have rigorously validated. Whether you prefer science or something else entirely for your belief system, one thing remains clear—the evidence certainly implies that epigenetic mechanisms have become instrumental in accelerating change on all levels of life. All inputs create a response and an outcome that lead to ongoing adaptive feedback loops. When leveraging precision data like genetics, you can create targeted strategies to inform dynamic change, which shifts the very fabric of consciousness.

Epigenetics: A Practical View

We can think of epigenetics as the switchboard for genetic expression. The primary genetic modulator that turns our gene expression up, down, on, and off like a dimmer switch. The changes to gene expression are caused by environmental influence rather than altered DNA sequence. But it's much more than just genes. It represents how we interact with our environment, covering everything from the food we eat, the water we drink, the quality of our sleep, how we deal with stress, the thoughts we think, and our emotional regulation. It extends to the environment, including the air we breathe, the cars we drive, the chemicals we clean with and use on our skin, and the location where we live. In short: pay attention to all aspects of life to truly make the most out of this invaluable science!

In recent years, the science of epigenetics has provided insights into how our genes are expressed. This information has led to a greater understanding of how we can influence our own health and well-being. Now, researchers are beginning to explore the role of epigenetics in consciousness. This emerging field holds promise for helping us better understand ourselves and improve our lives. We know epigenetics can affect how we think, feel, and act—and it provides powerful insights into how our consciousness works and what we can do to break free from old patterns that are holding us back.

Let's look at epigenetics and how it relates to consciousness, how genes and epigenetics can affect consciousness, and how scientists study epigenetics and consciousness. There is such potential with this research for the future of humanity and how you can start using these principles today to expand your horizons and dramatically transform yourself for the better.

Epigenetics as it Relates to Consciousness

It may seem odd to think of epigenetics and consciousness as related, but they actually have a very close relationship. Epigenetics is the study of gene expression changes caused by environmental factors, not any alteration in the underlying DNA sequence. This means that genetic changes can be passed down generations without any editing to the genetic code itself, resulting in alterations of physical characteristics and even behavioral traits. This research shows that our minds can actually shape our physical makeup via environmental influences, which could partly explain why some people are more emotionally or intellectually aware than others. So what does this all mean for consciousness? It could suggest, for example, that how we view the world around us affects our genes at a cellular level, thus influencing consciousness and behavior far beyond our immediate environment. Ultimately, epigenetics provides further evidence for the potential connection between mind, body, spirit, and behavior.

How Genes Shape Our Consciousness

Understanding how genes shape our consciousness is ever-changing and expanding as science progresses. Recent studies suggest that genetic makeup can influence myriad aspects of our experience, from overall temperament to the likelihood of experiencing mental or physical illnesses. While these findings can help inform better treatments and prevention for certain conditions, they also open up intriguing questions about the relationship between genes, environment, and nature versus nurture when formulating who we are as individuals. As technology continues to advance and new breakthroughs in gene editing enter the mainstream, it will be exciting to see where this research takes us next in exploring how much control we ultimately have over our destinies as a species.

How Scientists are Studying Epigenetics and Consciousness

Scientists are beginning a new era with the exploration of epigenetics and consciousness. Their fascinating studies focus on how genes, cultures, and experiences influence our behavior. Researchers are unlocking secrets about the way thoughts and feelings affect our bodies. They are discovering that we can "switch" certain genes on and off depending on our environment and lifestyle choices and understand how life events shape us. These leading-edge investigations can potentially revolutionize how science perceives heredity, evolution, and personal development. At this time in history, scientists have an especially exciting opportunity to learn more about how epigenetics affects the human mind.

Research on the Topic of Epigenetics and Consciousness

Recent epigenetics studies have revealed profound revelations about consciousness and its potential in humans. For instance, a research team at the University of Melbourne argued that epigenetics can explain how "trauma not only shapes a person's behavior but also influences those of future generations." Another research study by a group at Rutgers

University showed that experiences could affect an individual's behavior on a cellular level far beyond what we previously believed possible. Their research attempts to elucidate the underlying mechanisms between consciousness, brain networks, and epigenetic processes as they intertwine and shape our everyday lives. As more studies into epigenetics are conducted, it is exciting to see how closely consciousness is linked to our cells and even our ancestry.

Can Altered States of Consciousness Change Our Epigenetics?

Alterations of consciousness, such as meditation or psychedelics, are becoming increasingly popular as powerful self-exploration tools. Some suggest they can even cause change on the most basic level—epigenetics. Mind-altering techniques can influence gene expression, modify mood and emotions, and even affect physical health. In short, altered states of consciousness can potentially reset our epigenetic clock. Proponents say all transformations are possible with work and dedication, from unlocking blooming creativity to transcending mental illness. Many modern wellness experts see this as the next "big thing" in personal growth and development, allowing us to become masters of our own fate in a way never previously accessible. The preliminary data looks promising. Whether altering our consciousness can shape our epigenetics is still debatable, but you must admit it sounds pretty exciting!

Information: The Great Equalizer

Have you ever considered how much of a personal impact awareness can have on our lives and the world around us? It's astounding!

From our hard-wired DNA code to epigenetic programming, from biological markers in human systems to intangible feelings and beliefs that form a lens through which we see reality, information is truly the great equalizer. But just how aware are we of ourselves, our families, or even our culture? Are there gaps in our understanding of normal life stages for development within society? It may be time to take control by deepening our awareness surrounding these critical questions.

How much do we know about ourselves? How aware are we of the normal human developmental stages of life? Do we know that these developmental stages were transitioned and where they are at appropriate times or if our consciousness is arrested at a certain level of development?

My favorite map of consciousness is the Integral lens, specifically the STAGES model representing the life's work of Dr. Terry O Fallen, her brother, Kim Barta, and many others. STAGES is an integrally based development model. It charts human development from infancy to the highest levels of development humans are capable of. The STAGES model defines 12 stages in the evolution of human consciousness and meaning-making wisdom. Learning about these twelve general ways of experiencing and interpreting the world can help us understand our strengths, challenges, and unseen shadows. This model helps us to see where we are in the developmental stage of consciousness. It provides an opportunity for discovery, increased awareness, and actionable epigenetic strategies to become more aware of how we have become the individual we are today. Importantly, it can assist us in becoming the truth of what we are without our tremendous conditioning and inform our growth to contribute lovingly to our next personal expression of conscious evolution.

The Past Shapes the Future

Our past can shape us, but it doesn't have to define us. Is our story that someone or something has done something to us and they owe us? Instead of getting trapped in memories and feeling like something or someone owes us, why not choose instead to see the power we all hold to actively create our existence? Living in the present moment with no stress has profound potential: imagine being surrounded by a sense of peace instead of worrying about what's happened before or what could happen next. Achieving that kind of mindfulness might initially feel daunting, but even taking smaller steps towards living primarily "in the now" opens up endless possibilities for personal growth.

We can create an impact transcending past, present, and future generations. By being conscious of our beliefs, attitudes, and behaviors, we set a chain reaction that could determine what type of world lies ahead for those who come after us. Moreover, the information is recorded through epigenetics, so it lives on.

How do we flex this power? As a human systems designer, I guide human systems integration as the prime directive for creative expansion. Life becomes limitless when we are focused and aligned with our full expression. We each play an essential role in the direction of evolutionary expansion. Our beliefs, perceptions, thoughts, feelings, and intentions can be stored and transmitted to future generations through epigenetic mechanisms. So what are we willing to do to have a better life for ourselves and our future generations? Through action, promoting practices such as purposeful living leads to sustainability. This has been proven to make major changes over time when practiced collectively. Let's strive together towards lasting change that will positively shape ourselves now and define humanity forevermore! Here are a few shining possibilities.

Ancestral Transmission

Our ancestral heritage shapes us profoundly, influencing our physical and mental attributes and how we think and feel. Epigenetics expands what scientists understand about such inherited traits—not only can it reach back two generations to those who came before us, but it also extends further into the past to encompass even more of our ancestors' experiences. This is just one way epigenetics allows individuals across multiple generations to connect with their forebears beyond conventional boundaries—a powerful testament to the enduring nature of family bonds through time.

Our ancestral heritage informs our epigenetics from parent to child through their exposures, stage of consciousness, mental states and emotional experiences, socioeconomic status, as well as their cultural programming, and so much more. So it is no surprise that humans look, think, feel, and act like their parents and grandparents.

Familial inheritance is one of the most critical aspects of human identity. Epigenetics dramatically affects the future through transgenerational, multigenerational, and intergenerational transmission. Our grandmothers, mothers, and children all play a role in passing on epigenetic influences across generations. Whether it be the unique bacteria that colonizes newborn babies as they move through the birth canal or significant antibodies passed through breastfeeding. These act as a maternally derived immune system that protects the baby from disease and infections until it is old enough to create immunity. The antibodies in breast milk fend off invaders and control the newborn's intestinal homeostasis by dictating the diversity of the baby's microbiome. Transgenerational transmission has strong ties to both our past and future health!

Age Rejuvenation

How about the strategic use of epigenetics for age rejuvenation? We are in a time where we have access to modern technological advancements that will usher us into the truth of age reversal. We will begin to live longer, healthier, and happier lives. Strategies like apheresis (a technology that separates donated blood components to treat certain illnesses) with fresh frozen young plasma derived from college student donation have proven significant for age reversal. Genetic clock tests that show DNA age biomarkers can indicate how we compare chronologic age versus biological age, as well as intrinsic markers of our immune system strength, are now widely available through a simple saliva collection kit. Your biological age predicts health span (how healthy you are) and life span (how long you will live) more accurately than any previous molecular biomarker. It can be correlated to almost any health factor, such as physical fitness, socioeconomic status, and drug use history.

Imagine a time when many diseases that currently kill or create outstanding deficiencies will no longer be a thing. Imagine healing without the need for traditional medicine. A place where energy modalities combine both light and sound frequencies to treat illness. Daily we are discovering advanced ways to leverage these modalities to assist the human system in reaching a place of flourishing. Imagine a time when we do not get "sick." How different will life look when the word "healing" isn't even recognized or when the medicine is only reserved for complications of broken bones and accident-related injuries?

PTSD and Addiction

What about things like trauma and addiction? As a warrior sage, I lead a global movement for PTSD (post-traumatic stress disorder) resolution. My mission is to educate and restore wholeness to the human system following traumatizing military experiences. PTSD has been proposed to have a basis in epigenetic inheritance. It offers an excellent example of the relationship between physical and emotional well-being. Simply witnessing an extreme event can lead to severe unreconciled stress, which has been linked to decreased longevity and earlier onset of age-related diseases.

It is thought that the hereditary implications of trauma, addiction, and other immense emotional stresses are being encoded in our bodies through an epigenetic process.

At some point in the future, we will be able to identify an epigenotype for PTSD, addiction, and other epigenetic propensities, just as we now have LOCI reports for malignant breast cancers. For example, genome-wide association studies (GWAS) have successfully identified about 70 genomic loci associated with breast cancer. In genetics, a locus (plural loci) is a specific, fixed position on a chromosome where a particular gene or genetic marker is located. This epigenetic version of a genotype could guide us to a predisposition to drug and alcohol abuse, assist with precision interventions and reduce the risk of addiction in future generations.

Psychedelics, Empathogens, and Entheogens

Currently, we find ourselves in a renaissance of psychedelics, entheogens, and empathogens, which can loosen the intense grip of a programmed culture that lives predominantly in the separation of the mind. These molecules penetrate the central nervous system, and scientific and medical experts are just beginning to understand their effects on the brain and mind and their potential as therapeutics for mental illnesses when used with a trained guide.

Amazingly, they also provide a mechanism for us to know ourselves differently and to connect to our vast loving intelligence. These powerful tools, when used mindfully, can assist in taking down the walls of the Default Mode Network (DMN) and our sense of individuated self. DMN is an anatomically defined brain system that preferentially activates when individuals are not focused on the external environment. These psychedelics, entheogens, and empathogens can give rise to new states of consciousness that allow us to remove the veil between this current reality and other potential realities in the metaverse of creation.

When these molecules are done unintentionally by those seeking an escape from a life they do not enjoy, this process can be fraught with unintentional downstream effects. Hopefully, some sense-making of their use will be well constructed as we move into legalized medical and spiritual use of these molecules for trauma, couples' relational dynamics coaching, and consciousness growth.

Fortunately, our DNA blueprint and epigenetic propensities can guide more precise use of these life-changing tools. Additionally, knowing our energetic makeup and how our life experiences show up in the energetic system, which resides inside and outside the body, can give actionable information for who will receive a medicine well and who is best to avoid specific types of medicine (much like pharmacogenomics). Remember, we are all unique and will respond differently to these inputs based on that uniqueness.

Cultural Editors and Memes

It turns out we're all influenced by more than just our parents. Non-family members also shape who we are and influence how we live our lives and who we become. Psychologist Dr. Mario Martinez, author of the popular book, "*The Mind Body Code*," has referred to this as "cultural editors."

What other forces shape our worldview? While doctors, ministers, coaches, and friends have some impact, what about the greater systems such as democracy, religion, and government organizations? Every day we make choices that determine which paths will be taken in life. As a collective humanity, we also make multitudes of decisions every moment. So, let's examine who or what might sway those directions.

This untold influence brings us to another valuable point of exploration. From an epigenetic perspective, what role do memes play in shaping our worldview? Evolutionary biologist Richard Dawkins coined the word "meme" in his 1940 book, *Selfish Gene*, in which he compares the meme to the cultural equivalent of a gene. Memes are the phenomenon of ideas, behavior, and styles spreading from person to person within a culture. They can be a fashion craze, a process of how we do things, or the language we use to communicate. This memetic heritability can make it difficult to take hold of destiny's reins, but by implementing lifestyle strategies, you can create purposeful evolution more than ever!

Systems change requires robust planning and precise action, be it the human system, a corporation, governing body, or the greater environmental ecosystem. Cultural epigenetics combined with complex systems thinking provides a first-ever opportunity for wide-scale, global transformation.

Conscious Preconception Preparation

At Apeiron Center, our team is passionate about helping couples purposefully design their future and enjoy creating the family life they've always dreamed of. Through a comprehensive program, we empower potential parents with evidence-based strategies that create positive change across all areas of their lives. This program utilizes precision data to include the genetic blueprint, how epigenetics are expressed, advanced labs and other relevant health and aging biomarkers, beliefs, adverse and empowering life experiences, and levels of consciousness. So, when conception comes around, these soon-to-be families have everything they need to ensure lasting happiness and a high quality of life for themselves and their new addition.

Imagine a world where our new babies are so wanted that the parents consciously prepare. Feel what it will be like when these babies are parented with the awareness that they are connected to everything and taught to know love and connection. What will living in a world of love, compassion, and connectedness be like? We recognize that we are all one and that our development largely informs whether we know this. What happens when we embrace a world growing with unconditional love and understanding? How will it be when each generation is taught that their development shapes how they see themselves as part of something larger than themselves? It's not just possible but powerful to embrace who we truly are—infinite beings connected through limitless potential!

Precision Performance Medicine (PPM)

Precision Performance Medicine (PPM) is a new field representing a new health paradigm. Contemporary medicine focuses on treating illness and preventing disease, whereas PPM focuses on utilizing our medical technology and advancements to enhance human function and performance. It is a strategic process that involves evaluating all aspects of the "human system," body, mind, and spirit, taking into account each individual's physiology, genetics, genomics, and their expressed need to create evidence-based therapies to promote youthful longevity and peak performance.

As many more novel, leading-edge strategies come on the scene. These revolutionary methods will inform our ability to evolve our human lives purposefully and intentionally. These opportunities will see a shift in the planet's consciousness, where everyone has the potential to have their basic needs met. To live safely, be treated with dignity and honor, and know we are all valuable contributors to each other, the planet, and all creation—cheers to the life-changing science of epigenetics.

Making Change When Change is Hard

by Abby Kreitler Hand, MSN

Lasting change is hard. You don't have to do it alone.

Abby Kreitler Hand

At 37 weeks pregnant in November 2020, I found myself lying upside down on an ironing board propped against the couch. The next day I was due at the hospital for an External Cephalic Version (ECV) to, fingers crossed, flip my stubborn baby out of the breech position. The previous attempt, lying on a massage table in my midwife's carport, was unsuccessful. It felt like my skin was tearing as the two midwives attempted to maneuver the body inside my own. So, we called in the big guns. We had come so far, baby and me, and we weren't going to give up on our plan to deliver at the birthing center. Not when we had successfully evaded Covid for so long and still had unused tools in our toolbox. But I hoped these holistic interventions would encourage my baby to move into a safe position for a natural birth. So, in addition to inversions, my routine included moxibustion, acupuncture, pelvic tilts, visualization, meditations, forward folds off the couch, shining a flashlight and playing music between my legs, and so on. If it was in one of my crunchy mama books, I tried it.

On this last day before the ECV, staring up at the line where the wall and ceiling meet, I had two thoughts: 1) I accept there are things outside my control, but damn if I won't try to control the things I can, and 2) I have unknowingly spent the last three years training myself, mentally and physically, to begin this wonderfully insane transition to motherhood. Surprisingly, I felt an intense sense of gratitude for my autoimmune disease, the way it allowed me to grow, and the skills it forced me to learn— quieting my mind, nourishing all aspects of my being, and genuinely listening to my body. Lupus could have remained a black shadow over my health for the rest of my life. Instead, I created something positive from my experience by serving as an epigenetic coach and health consultant for prenatal clients.

Resistance

This moment in time was certainly a highlight, of which there have thankfully been many; however, there have been definite lows along this autoimmune journey as well. I constantly bailed on my friends and family due to exhaustion, denial of the butterfly rash that spread across my face, fear of going to yoga because my GI system would embarrass me, and so on. The most traumatic, however, I named "The Big Event." One morning, many years before my transformation, I started having trouble with my vision when I went about my work as a medical assistant at a busy clinic. I blinked my eyes, and it resolved, so I paid it no mind. Later in the afternoon, the blurriness returned, this time accompanied by bright sparkles on the edges of my vision. "I've been staring at my screen too long," I thought, "I should take a break and get some air." The thoughts were lost as a gargantuan pile of charts was dropped on my desk, and I buried myself back into work. The blur and the sparkles returned as soon as I got to work the next morning, this time with a squiggly worm-looking thing blocking half the vision in my right eye. "What fresh hell is this?" I pondered and immediately noticed the fingers on my left hand begin to tingle, that sharp feeling as if my hand had been asleep. I opened my mouth to tell my coworker about the strange sensation, but it felt heavy and numb like I had just smashed an entire pint of ice cream. Abruptly the world around me began to dissolve. Panicking, I headed to an exam room to lie down. My friend followed me in, concern in her voice and probably on her face, but I couldn't see it. My vision had gone completely black, and the tingling now turned to numbness, strangled my entire arm. I couldn't see or feel my arm and was slurring my words.

I was terrified. "Holy shit, I'm having a stroke. Please don't let me die." My thoughts raced. "What the f**k. I'm only 26." I was immediately sent to the hospital for an MRI, which, thankfully (and confusingly), came back normal. While I was ecstatic to a) not have had a stroke and b) not be dead, I was left reeling without any answers.

So how did I get from laying on an exam table, fearing for my life, to laying upside down on an ironing board expressing gratitude for this ridiculous autoimmune disease? Access to my genetic code, incredible support, and a lot of hard work. It was a long road.

Following "The Big Event," I sought help from a compassionate neurologist who eased my fears by explaining that I had likely experienced an atypical form of migraine and recommended I start a medication to prevent future episodes. I felt instant relief and hope. I could handle this! But then Dr. R mentioned my other symptoms and started throwing around words like Multiple Sclerosis, Myasthenia Gravis, and Systemic Lupus Erythematosus. I was floored. Dr. R referred me to a rheumatologist named Dr. B, who began the long and tedious process of autoimmune diagnosis. Sidenote: Did you know the average time to receive an autoimmune diagnosis is roughly four and a half years? Not only that, but an individual may see up to, on average, 4.8 different doctors before they get any answers. I fell squarely into both these statistics. Several years earlier, at 23, my Primary Care Provider determined that my fatigue resulted from depression and put me on a battery of SSRIs, SNRIs, and other anti-depressants, none of which had any positive effects (though all had abundant negative results). This marked the true beginning of my lupus journey and reinforced an already tenuous relationship with conventional medicine. Sixteen vials of blood and thousands of dollars worth of tests later (thanks, health insurance!), Dr. B determined I had Undifferentiated Connective Tissue Disorder (UCTD) and started me on hydroxychloroquine, the gold standard for many autoimmune disorders.

The rub here is that UCTD is more of a state of limbo than an actual diagnosis. It's a placeholder that says something brewing, but we don't know exactly what it will morph into. In the meantime, here are progressively more toxic drugs whose side effects are just as harmful as the symptoms we try to manage! This could have and should have been the moment when at least one of my three doctors intervened and encouraged me to adopt a lifestyle that better supported my health. Even then, we knew the incredibly positive effect proper nourishment, movement, and stress management could have on autoimmune disease. But instead, they jumped straight to a drug to manage the symptoms of autoimmune disease (hydroxychloroquine), a drug to manage heart palpitations and tachycardia (propranolol), a drug to prevent migraines (topiramate), and a prescription-strength non-steroidal-anti-inflammatory (NSAID) to manage joint pain (meloxicam).

The hydroxychloroquine also saddled me with frequent visits to the ophthalmologist to monitor my vision because this drug has a high risk for retinal damage with increased time and dose. In addition, the meloxicam further wrecked my stomach, and the topiramate altered my taste buds, making my food taste sad, which then made me sad. Finally, the propranolol mellowed me out to the extent that a doctor I worked with remarked that maybe the propranolol was *too* effective. In other words, my productivity was more valuable than my health.

Though I did start feeling slightly more energetic and less stressed, thanks to the numbing effect of the propranolol, I still lacked balance, and my providers came up disappointingly empty in this area. There was no mention of sleep, nutrition, movement, or even an assessment of cigarette and alcohol use, both of which were occasional social visitors at that time in my life. Small changes in these areas early in the game could have significantly impacted my path, but unfortunately, I had not yet gained these insights. I was still relying on others to navigate this health crisis for me.

On top of the vague, limbo-like autoimmune diagnosis, I was suffering from Passenger Syndrome. What is Passenger Syndrome, you ask? It's the psychological condition of being passive in your health journey instead of an active participant. Symptoms may include not understanding your diagnosis, treatment, and/or long-term plan; blindly accepting your medical provider's guidance as the only way forward; feeling like you can't ask questions for fear of being seen as dumb or difficult; and/or sticking with a doctor even though they are patronizing and have terrible bedside manner. After years of self-discovery and self-experimentation, I eventually learned to manage both conditions (lupus and Passenger Syndrome). Still, at this point in the story, I had a long way to go before reclaiming my power. Back then, I worked a stressful job, dutifully swallowing my prescription drugs, and ignored my body's urgent cries for help.

Two years passed, marked by urine samples, blood levels, and field of vision tests, none of which were explained in detail. Yet, I powered through, as we are conditioned to do. I found an amazing partner on a journey with his own autoimmune disease. I began an accelerated nursing program, then received my Master of Science in Nursing. Five years after my initial complaints of fatigue and discomfort fell on the deaf ears of my PCP, at twenty-eight, I finally received a diagnosis of Systemic Lupus Erythematosus (SLE). It was a bittersweet day. I felt torn between gratitude for finally having an answer, anger at how my body failed me, and frustration with the lackluster medical care I had received. This could have been another tremendous educational moment; however, I was provided with no additional information, guidance, or options for handling this news.

At thirty, I began working nights as a labor and delivery nurse, and it didn't take long to see it was a poor fit. I kept a nocturnal schedule, even on the nights I didn't work, to prevent continually messing with my sleep and subsequently triggering an autoimmune flare. This was fun for a while, but it's isolating and discomforting to be awake when the rest of the world is asleep. I didn't feel adequately trained to deal with the complexity of my patients, and my unit was straight out of the movie Mean Girls! Worst of all, the overall hospital culture did not align with my values, particularly in treating patients who didn't fit into the white, upper-middle-class demographic. The last straw for me was that they didn't want to give me time off for my wedding! I gave notice and married my handsome Mr. Hand a month later at a tiny chapel in the woods of upstate New York.

We accepted a temporary assignment in Singapore for my husband's job the weekend of our wedding, and within a month, we moved abroad. This turned out to be an exciting and challenging experience. Based on my visa status, I couldn't work, and transferring my nursing license for such a short time didn't make sense. So, I spent much of my time traveling solo throughout Southeast Asia, immersing myself in history, culture, and food. It was an amazing opportunity for which I am incredibly grateful. Unfortunately, Mr. Hand worked crazy hours, and I had few social outlets. I didn't know it then, but I was incredibly isolated and likely depressed. I gained all the weight back that I had lost for the wedding (perpetuating a cycle of yo-yo weight loss), felt my autoimmune symptoms explode, developed intense sugar cravings, and was unhappy with how I looked and felt. I'd had it, and I needed to make a choice. I could continue fighting my body or finally stop and listen. Luckily, my husband helped me strategize a way forward and supported me without judgment. He introduced me to new ways of approaching health and sparked my passion for learning and personal growth via personalities like Tim Ferris, Rhonda Patrick, and Jocko Willink.

Acceptance

When we returned home to Texas, I was on a healthy living warpath. For the first time since childhood, I could listen to my body physically, mentally, and emotionally. I rejected the notion of what a nurse is "supposed" to be and accepted an outpatient RN position at a Functional Medicine practice. I hoped to absorb information to heal myself while also helping others. And like so many who first stepped into the alternative medicine space, I was drinking from a fire hose.

I didn't know where to start with so much information and many options. This resulted in a shotgun approach to lupus management, trying everything all at once. One of these interventions had to work! I put my faith and money into supplements, sometimes taking handfuls at a time. I dipped my toes in the Wahl's Protocol, Paleo, and the Ketogenic Diet, completed several rounds of Prolon, agreed to a battery of specialty lab tests, bought a sauna, began a daily meditation practice, started an intense fasting regimen, joined a fancy gym, started Low Dose Naltrexone (LDN), gave myself weekly B12 and glutathione injections, and even experienced a few guided ketamine sessions. Altogether these interventions had a very positive impact. My energy improved, and joint pain and swelling decreased. Even more exciting was that I hadn't experienced a twinge of migraine pain since starting this path. That was huge, but it didn't last long. I was doing too much all at once, and as the excitement began to wane, the stress of trying to manage it all had me approaching mental burnout.

On top of that, my most valuable metric, the lupus markers, was not improving. I began to question the efficacy of this approach. Don't get me wrong. These are fabulous tools given the appropriate time, place, and follow-up. In my case, however, I didn't know which interventions were helping and which could be hurting. The idea of testing one thing at a time felt tedious and exhausting. Thankfully, on the edge of scrapping it all, I went to a health conference and was introduced to the world of genetic testing and epigenetic coaching. This is where I began to soar.

Surprise

I signed up for an epigenetic coaching program and started the deep dive into genetics as a blueprint for health optimization. I ran my data and was shocked to find that some of the choices I believed to be healthy could increase inflammation, making managing my autoimmune disease more challenging. This was incredibly frustrating, and while I could have focused on the wasted time and money, I chose to view the work leading up to this point as an important learning opportunity. Besides, I now had my genetic map, which acted as an instruction manual. I was ready to start a more streamlined course of action. But where to start? No matter what genetic company you choose, the information can be overwhelming, and it was hard to resist that same shotgun approach from before. Thankfully, in addition to this incredible information, I also gained a supportive community that helped me understand the benefit of taking a more measured, scientific approach to lifestyle and behavioral change. I no longer felt like an island. This is one reason it's so important to work with a coach, or another experienced guide, to help create an actionable, efficient, and realistic plan. We've already fallen down the rabbit holes, so you don't have to. That said, transformation and health optimization will always require a healthy dose of self-education and personal exploration.

Taking it slow this time, I assessed nutrition first, and my genetic report supported many of the changes I had already made. However, several of the choices I assumed would be beneficial needed to be in alignment with my innate ancestral intelligence. So I set out to design a diet guided by my genes, food preferences, and what was realistic for my life. An unexpected benefit from this plan was that it allowed my focus to shift away from weight management and toward nutrition as a tool to reduce inflammation while also promoting nourishment. It also helped emphasize what I could add to my diet and not just fixate on what I chose to remove. This way, I found a framework for my nutrition that wasn't restrictive. One fascinating recommendation I received was to reduce the number of nuts in my diet. Nuts are great for many people; they contain vital nutrients and are well-supported in the literature for their health-promoting benefits. Unfortunately, they are also high in polyunsaturated fats (PUFAs), which, combined with my genetic makeup, have the propensity to cause increased levels of inflammation. Not helpful when you are trying to combat autoimmune disease, which is inherently a disease of inflammation. When I cut back on nuts and opted for those higher in monounsaturated fats (MUFAs), like macadamia nuts, I felt a positive shift in my digestion and improved joint discomfort. Over the years, I have seen the same effect in clients with a similar genetic picture, each one following the same sequence: resistance, acceptance, surprise, and then gratitude.

Another example from my genetic report is a fun gene variation, or single nucleotide polymorphism (SNP), nicknamed the "cookie jar" gene. People with this variation can't have just one; the whole box is almost gone before we realize it. This is related to the gene's ability to reduce inhibition while simultaneously increasing desire. This information helped me improve my awareness and find ways to set myself up for success. It allowed me to see that this was not an issue of willpower; I am not a failure. I could write for days about the fascinating ways genetics helped me understand the deep connections between genes, neurochemistry, and choice. However, you must experience it because people are unique, and your genetic picture may look very different from mine or your close family members'.

The second stage of my health overhaul focused on micronutrients. When I started experimenting with nutrition, I also cut out the handfuls of supplements apart from those I considered truly foundational. Due to modern agricultural practices, obtaining our vital nutrients through diet alone has become increasingly difficult, and supplementing high-quality products can close the gap. Unfortunately, with the emphasis on *quality*, which is not always accessible or realistic, we fortify many common foods with synthetically derived nutrients. While necessary and effective at a population level, this approach may negatively affect certain individuals. Consider this scenario: Folate, also known as vitamin B9, is a micronutrient crucial for fetal brain development and preventing neural tube defects. And while the public health campaign to fortify foods with folic acid, a synthetic version of folate, has been wildly successful, it doesn't consider individual risk. Due to variations in genetic programming, people need more tools to convert folic acid into a usable form. If supplementing with folic acid, you may not adequately correct the folate deficiency and may inadvertently create excess. This can lead to poor outcomes for the individual or, in the case of pregnancy, the developing child. So, should we then tell everyone to take methyl folate, an activated form of folate? Probably not. Some people's genes put them at risk for over methylation, causing anxiety, agitation, and exacerbation of obsessive tendencies. To me, the complexity of this picture drives home the need to give recommendations at the individual level, if reasonable, instead of basing them solely on population-level evidence. As technology advances and becomes more affordable, I think it's reasonable to see genetics find its place in mainstream medicine and help address these issues.

Genetically speaking, I am in the category of people who should be cautious with folic acid. I also discovered I likely had decreased function in an enzyme required to convert inactive vitamin A into the active, more usable form. Carrots are not going to cut it for me. I may also need higher amounts of certain B vitamins. I discovered I need to be cautious with supplements like zinc and vitamin E and found ways to support my Phase I and II detoxification pathways. This data helped me tailor my nutrition further and allowed me to focus on food as medicine instead of resuming the handfuls of supplements. With the help of my favorite tracking app, Cronometer, I experimented to see if I could reach specific micronutrient targets through diet alone. I upped my leafy green vegetables for folate and vitamin C, ate one Brazil nut a day to improve selenium levels, and continued tweaking my diet in these tiny, subtle ways. Even with such a purposeful and intentional way of eating, my nutrition still fell short on several targets. Knowing that autoimmune remission relies heavily on meeting these vital requirements, I resumed a few key supplements. This time, instead of falling prey to health trends and marketing, I had better information and felt empowered to make more informed choices. I carefully curated products using ingredients that matched my genetic needs and added them back one at a time to examine their effect. I relied heavily on my tracking app for this phase and blood markers, which are another powerful tool, especially when combined with genetics.

These first two phases had their difficulties, yet nothing compared to the challenge of aligning my sleep behaviors with my sleep-specific genes. Up to this point, I had altered behaviors to influence genetic expression positively. You can't do that with sleep. Our sleep genes are ancient. We share them with insects and cyanobacteria. You can choose to work with your genes or fight against them, and the latter is a very difficult road. Had I known this at the beginning of my journey, I would have prioritized sleep and likely seen improvements much faster with an overall better quality of life.

One of the most important concepts I learned through my epigenetic training is that, although you may think you can adapt to less sleep, you're accepting a decreased mental and physical performance level. You're choosing mediocrity. According to my genes, I should go to bed between 8-10 PM and wake up between 4-6 AM. No way that was going to work! Why couldn't I have the rare DEC2 gene that allows people to function on significantly less sleep? I fought it for a while. Luckily, my wonderful husband had already headed down that road, so besides being an accountability partner, I also benefited from his positive role modeling. The world was much easier to walk through once I joined him.

Regardless of your genetic propensity, you're likely not getting enough sleep, or at least not enough quality sleep. It's a social epidemic. So, in addition to sleep timing, I worked on my sleep environment, hygiene, and tracking. These are highly impactful tools that anyone can employ with guidance from a health coach or a little research.

This epigenetic transformation's final and most daunting process was breaking up with chemical-laden body care and household products. Regardless of your genetic propensities, avoiding endocrine-disrupting chemicals (EDCs), known carcinogens, and other ubiquitous chemical compounds can positively impact your long-term health. Almost four years later, this remains a considerable challenge. Just like with nutrition, product labels are tricky and often misleading. Yet it is so worth doing the work, and given my genetic picture, it is incredibly important for both my lifespan and my health span. What's the point in living a long time if my quality of life is in the gutter? Including health span, not just lifespan, in our health conversations is a must. Environmental exposure is a large part of that conversation and doesn't get enough airtime. Yes, our bodies are beautifully designed to metabolize and eliminate the substances we encounter daily, but we need to support these processes intentionally. Otherwise, we're setting ourselves up for chronic disease, hormone imbalance, cancer, etc. For me, this is done through high-quality sleep, real food, sweat, and other, more specialized tools. Things I wouldn't have connected without the help of my genetic code and an understanding of complex systems.

As I dug further into the research surrounding my genetic variations, tweaking and fine-tuning my experiments became second nature and enjoyable. In addition to seeing considerable improvements in my physical and mental health, building health competency gave me the confidence to stand up to Passenger Syndrome. I fired Dr. B and came on board with a new rheumatologist who was a complete outlier from my other providers. To my surprise and delight, Dr. P asked whether I had ever considered coming off my medications. My eyes welled. I felt heard and validated for the first time since starting this journey. I spent five minutes with this man, who recommended the thing I had badgered previous doctors about for years. Even if the experiment didn't work, at least we were acting instead of resignedly accepting a slow march to death. With his help, I came off both hydroxychloroquine and the LDN. He was also the first to question if my birth control was creating an unnecessary need for blood pressure medication. This hypothesis turned out to be correct, and with the support of my Women's Health Nurse Practitioner, I knocked two more medications off my list. The remaining prescriptions, topiramate and meloxicam had already fallen away thanks to improvements in my joint and migraine symptoms. I was prescription free! Lab testing and blood pressure monitoring four months later confirmed that I was in lupus remission without medications, and my blood pressure was back in a healthy range.

I have a picture from this day. It's of my childhood friend and me celebrating twenty years of friendship and our successful remission from an autoimmune disease. To make it even more magical, we were at a Dashboard Confessional concert in the middle of a rare Texas snow. Everything about this photo shouts joy, relief, and triumph. It was an incredibly emotional and radiantly joyous day, one that I hold up to the light when life begins to darken around the edges. A reminder and beacon for when I lose sight of my goals or when doing the work feels too heavy. Because it is a lot of work, it takes mountains of patience and perseverance. But it's worth it. About six weeks after bopping around to nostalgic emo music, I learned I was pregnant. It was an anxious time, not only because it was the first week of the Covid-19 lockdown (I ordered pregnancy tests from Amazon because I didn't know if we were allowed to go to the pharmacy), but on top of that, we weren't actively trying to conceive. Almost 50% of all pregnancies in the US and globally are unplanned. This speaks to the benefit of both men and women, capable of reproducing, to live a healthy and intentional lifestyle. It's also why I'm so passionate about preconception health consulting. Everyone benefits.

There was so much unknown at that point; however, I was confident that I could handle whatever challenge Covid or pregnancy, or the combination of the two, decided to throw my way. I was the healthiest I'd been in my entire life, and I had my genetic blueprint spelled out exactly how I needed to support myself during pregnancy and establish a strong foundation for my child's lifelong health. It's hard to know precisely what would have happened if autoimmune disease still dominated my life. The thought of a lupus flare during pregnancy gave me nightmares. However, I can confirm that having lupus gave me the space to prioritize health and shift my mindset. It also inspired me to create a professional path dedicated to building a healthier world.

Gratitude

It was, and continues to be, a wild and challenging journey, and I've learned so much about health and personal responsibility because of it. The hospital ECV was successful and surprisingly easy, thanks to modern obstetric medication. With that obstacle out of the way, I spent the next three weeks preparing for the hippie-dippie natural birth of my dreams. But just when you think you have everything figured out, the Universe will send a new learning opportunity. Three days past my due date, I ended up in the hospital for induction. After a wild couple of hours of labor, my son came crashing into this world with a head full of blonde hair and the longest umbilical my medical team had ever seen. He was perfect from the start, as is his birth story. It was an incredibly healing experience that helped mend a rift within me and find a more balanced and grounded perspective on women's reproductive care. It also opened space for me to become a stronger and more centered guide for folks on the path toward parenthood.

As you may have noticed, many of the brilliant women involved in this collaboration were drawn to health optimization, particularly epigenetics, because of their unique health journeys. They battled debilitating symptoms, psychological trauma, unending roadblocks, internal (and external) doubts, and crises of confidence. They were either over- or underwhelmed by the treatment options presented, or sometimes both. But they also found joy, validation, strength, fellowship, and sovereignty by stepping outside the conventional approach to medicine. They listened to their bodies and committed to doing the hard work. Now they seek to guide others through this magical and often painful process of self-discovery and transformation. I'm no exception. My story contains all these twists and turns, and I'm sure yours does, too.

We are in an epidemic of exhaustion and burnout. We are living through an explosion of chronic health issues, crumbling environmental health, and an ever-increasing cultural divide that permeates the world of health. Everybody has something to say, but no one takes the time to listen, and this might be the single most significant and most easily preventable cause of suffering in healthcare. Our current medical model is not designed to listen. We may hear but are not listening, which extends to ourselves. We are not listening to our bodies. Our natural and innate cues are being ignored, replaced by personal and societal pressures, glowing screens, and access to various foods engineered to perpetuate addiction. We are not listening to ourselves or each other, making us sick.

As a coach and consultant dedicated to helping others navigate their health and well-being, my strongest tool is the ability to listen not only with my ears but also with my heart. Coaching comes from compassion, empathy, non-judgment, and openness because I'm on a journey, just as you are. I may not always have the answers, and I can't climb the mountain for you, but I can walk the path beside you. My second strongest tool is genetics. Having your genetic data is not a requirement for pursuing optimal health, but it does provide a pretty stellar blueprint to help guide your actions and success. My clients are excited about genetics and this radical approach to health management and family planning. They embrace the paradigm in which the "responsibility" of a successful pregnancy and a healthy child no longer falls solely on the woman. They understand that, in addition to the mother's health, the father's health during the preconception stage impacts the child's long-term health. Genetics is a gift that allows parents a glimpse of their internal programming. It guides them towards optimal preconception health, improved fertility, healthier pregnancies with fewer complications, improved postpartum recovery, and, most importantly, healthy foundations for future generations. We are creating a path to a more resilient future, and I hope you join us.

Genetic Clarity Through Courageous Action My Path to Transformation

by Laurie Kaplan, RDN, LD

To be courageous is to stay close to the way we are made.

David Whyte, poet

While this chapter is not an essay on courage, it is one of living true to oneself and what it takes to get there through understanding the wisdom of our DNA. Living according to our genetic blueprint provides the foundation to lead and enjoy a courageous life. The French philosopher Camus used to tell himself quietly, "to *live to the point of tears,* not as a call for maudlin sentimentality, but as an invitation to the deep privilege of belonging and the way belonging affects us, shapes us and breaks our heart at a fundamental level." Albert Camus and I share a similar thought regarding living such a life. In my mind, living from your heart and with your heart is what we strive for in this experience of being human. I believe living from your heart is the safest way forward to one's truth.

I was dying.

I was dying in a relationship—a marriage to a person I thought was my soulmate. Disregarding what my mind and body truly needed to thrive and thus show up and contribute to this world, I was in survival mode choosing sub-optimal foods and drinks, thoughts, and habits. Disrespecting my biology, physiology, and psychology, I knew deep within me that if I did not make significant changes and fast, I might not be around to be the mother, friend, sister, and daughter – the *human I* knew myself to be. Always curious about my health, mind, and body from a scientific level), I started to hear about "biohacking."

Ben Greenfield, Tim Ferris, and so many others caught my attention. How can I *really* feel better deep within my bones and soul? Can their teaching and guidance and all the n-of-1 experiments help me find the answers for my body and mind? Do I need to soak in ice baths and take 50 supplements daily? Only eat kale (or not kale!) and drink hydrogenated water? I do not mean to make light of these experts or their education of many supportive products, including ice baths and other forms of cold therapy. Many of them are valid and can be highly supportive of health optimization. I am a Kion certified health coach, which was Ben Greenfield's foray into coach training development. And thanks to Mr. Greenfield, I chose to get my DNA tested. Thus, I began the deep dive into my genetics and the wisdom within.

My genetics indicated I should be overweight, have difficulty losing weight, be at high risk for diabetes and cardiovascular disease, and have a powerful genetic propensity to handle stress poorly. Genetically, I also have the ability to build muscle and be very strong. This hit home like the mace I just purchased from Onnit.com. Uh, yeah. No wonder I felt like I was dying, living in an environment of psychological abuse (no blame here) year after year. I have a genetic SNP (single nucleotide polymorphism - more on this later) that prevents the clearance of the neurochemicals that influence stress. Mainly norepinephrine and dopamine. I am what we call a "worrier." Yep. So was my dad. While I am unaware of his exact genetic expression, I know that he certainly showed up as a heavy-duty worrier and would put money on the fact that this gene expressed the same as mine. So, living in a world that even the most genetically chill person would find difficult, this gene (and others, as we are polygenic beings) was driving me crazy.

Discovering myself from a genetic level was the first step in learning to truly be me. It became key in my health and wellness optimization journey to *live* with my genetic variances and quirks, one that I was ready to step into with all my courage.

I hold two science degrees from Michigan State University - a BS in Animal Science with an Equine emphasis and a second BS in Human Nutrition. I completed my dietetic internship (required to become a Registered Dietitian) at the University of Kentucky with the idea that I could be a nutrition coach for top-level equestrian teams. Having been involved with horses since age five and working at various equestrian establishments, this would be the perfect blend of my skills. However, my life quickly took a different path when I joined the United States Air Force, where I met my husband at officer boot camp. I briefly worked as a clinical dietitian in a hospital setting. While I enjoyed working, I was frustrated with some of the limitations I could already see in clinical dietetics and realized I did not want to return to this form of nutrition practice. Therefore, I happily chose to stay home to raise our two amazing boys, Alex and Christopher. As they grew up, I gradually felt the pull to use my training again to support others on their health and wellness journey. When I realized I needed to transition out of the marriage, I returned to my education earnestly, knowing I wanted to use my nutrition knowledge and teaching skills to help others. Listening to a podcast from FDN (Functional Diagnostic Nutrition) and Reed Davis, I heard Dr. Melissa Petersen discuss the world of epigenetics, nutrigenomics, and nutrigenetics. The proverbial light bulb exploded in my head. THIS made sense! I knew THIS is what I wanted to pursue; to provide clarity to others in the insanely confusing and noisy world of health and wellness optimization. Dr. Melissa was part of the amazing Aperion Zoh group at the time, and I attended their in-person training in Chicago, a couple of hours from my home in central Illinois.

I was nervous about attending the workshop. Foremost, I wondered if I was smart enough to be in the same room with these folks. Brilliant minds, including the founders of the Apeiron Zoh group, Dr. Mickra Hamilton and Dr. Dan Stickler, and the other coaches and trainers, including Dr. Melissa Petersen, Meaghan Foley, and Daniel Luper, would be there presenting the training. Would they think I wasn't smart enough or had been out of the nutrition science field for nearly 20 years? For many years, to stay safe and alive, I thought I needed to silence "me and my voice." So, the thought of speaking up, asking questions, and sharing ideas made me feel like a horse seeing a flying plastic bag whip around her pasture, threatening to eat her alive.

What I discovered was the feeling of a new home, where I received a warm, welcoming embrace. These highly intelligent individuals were excited to look at the beauty of being human through the lens of genetics, and I felt inspired to learn from them and join the team. This consolidated my mission and educational focus, using genetic code wisdom and creating a best practice guide for others to learn. Emphasizing imperfect beauty can be a stage from which precision human longevity can be optimized.

While I considered myself quite independent, the thought of leaving the marriage without a means to support myself was terrifying. Becoming a certified Epigenetic Health Coach with Apeiron allowed me to create my own coaching program. I also found clarity for my next life stage by learning how these lifestyle genes express themselves. This provides the myriad of ways coaches can lead their clients to live the life they seek. My desire to assist others in clarifying their genetic road map while cutting through the chaff of health goal planning is what I felt called to learn.

Why? The odds of being born are insanely small—something like 1 in 10 to the 2,685,000 power. Basically, "zero," according to Dr. Ali Binazir. So, listening to the voice inside me saying, "Please, go do you and be

YOU so you can help others," was starting to feel like a very significant why— something to pay attention to. Now I had my why and my how. Without hesitation, I signed up for the coaching academy with Aperion Zoh and have not looked back.

Then the fun began. I received the results of my genetic testing:

> ADIPOQ (2) GG: Risk of increased appetite
>
> COMT AA: Reduced or slow clearance of neurotransmitters, reduced estrogen metabolite handling (possible cancer risk?), increased sensitivity to stress, good memory, and attention to detail.
>
> DRD2(1) GG: Strong addiction propensity.
>
> LEPR (2) GA: Increased leptin resistance probability and increased hunger.
>
> SLCA2 GA: The "sweet tooth gene."
>
> PPARGC1A CT: reduced mitochondrial (energy) biogenesis, increased need for regular exercise.
>
> NR3C1(3) CC: Increased cortisol response to stress, increased anticipatory cortisol response.
>
> PAI: Increased risk of stroke
>
> ACVR1B AA: enhanced power and strength

What the heck does this mean? When one of my clients first saw her genetic report with similar "code" words, she said it looked like hieroglyphics. Within these codes, this language of humanity lies our bodies' potential limitations and unlimited potential. So, learning to "read" genetic code became my mission for the next few years and continues presently (as more and more research is released almost daily).

Upon passing the exam for the extensive Apeiron epigenetic health coach training program that provided deep genetic and epigenetic science in the areas of sleep, nutrition, detoxification, stress response, hormones, and others, I started to gain an understanding of how to translate this new language for understanding my genetic makeup and guiding future clients to their genetic clarity.

As there is much crosstalk between the genes, looking at just a few gene variants does not provide the whole picture of the person. This is because we are polygenic (meaning "many genes") beings, and when we consider utilizing the genome to learn what our body can thrive on, we must learn and consider how these genes work together. A glorious and sometimes cacophonous symphony of cellular information which can even come with GPS!

The abbreviations above were once, and sometimes still are, referred to as genetic mutations. I prefer the description of a typo or misspelled word. Words can create a heaviness to which our body can respond in subpar ways. Explaining to clients that they have a "mutation" can set the stage for guilt or feeling "bad" about their bodies. We certainly do not need any more shame in our lives, so precision language is vital to my coaching. By understanding how each gene is "spelled," we can dive in and design a health and lifestyle plan that supports the ideal expression of those genes.

For example, I have several misspelled genes or SNPs (single nucleotide polymorphisms) that indicate I should be hungry, find it difficult to lose weight, battle addictions, crave sweets, and be challenged when it comes to handling stress. While some of those issues do express for me, namely the stress piece (that set of genes can be highly influential and tricky to manage), I have never been overweight, felt excessive hunger, and can lose weight easily.

I mentioned the weight and power that words can have. When coaching clients, I explain that there are no "bad" genes. If you have ever studied or considered NLP - neurolinguistic programming, you can appreciate how powerful words and concepts influence our moods and choices. While some genes can create more metabolic mischief and certainly real health concerns when not managed properly, these genes can also have beneficial expressions. For example, when you look at the SNP COMT (Catechol-O-methyltransferase) - the Worrier vs. Warrior gene (yes, my SNP expresses as the "worrier" as I referenced previously), my variant also indicates that I have a strong memory and good attention to detail—the proverbial two sides of a coin.

Epigenetics: "Epi" from above, "Genetics" the Genome

Epigenetics describes the ability to alter our genes' expression for health optimization. We have the power to make choices that can turn specific genes off or on - increasing or decreasing their expression. How can I quiet the expression of that pesky Worrier SNP I mentioned before? For me, it's reducing harmful stress and focusing on living my truth along with proper (and delicious!) nutrition, activity, and targeted precision supplementation at certain times. How do I stay out of a fight or flight mode where this gene stays "on?" I tap into the activities that make me feel like I am home, working on my farm. Using my body, nourishing others with food and fun, connecting with clients, playing the drums, and playing with my horses are activities I connect with that ground me.

My genes indicated I should be an overweight, starving person with diabetes and cardiovascular disease, and difficulty with addictions. So, why had I escaped this? Being involved with horses since the age of five has kept me relatively healthy for much of my life, thus maintaining an ideal weight and, for the most part, good blood/metabolic lab values. However, my weight and blood markers took a deep dive off into the land of very unhealthy when I was not living ME. I was so mentally and physically misaligned that I felt I was facing an early death. Understanding my genetics has given me the most valuable information.

I have a set of genes that indicate I can develop elite-level strength and power. Despite being rather petite (5 foot, 100 lbs.), I have worked in horse barns and farms (including my own, where I currently reside) and have generally felt confident getting the heavy stuff done. Ok, sometimes a taller person may need to help lift a 50 lb. bale of hay into the stack... we all need a little help sometimes.

Now that I am back working on my farm, I continue to flourish physically and mentally. I am in the place my genes want me to be. Taking care of my land and animals helps quiet the worrier. KNOWING that I can express myself as a worrier does not scare me - it gives me the power and the ability to be radically responsible about how I want to show up. My horses don't want to be around me when that gene is expressed, as they can sense my anxiety.

Understanding how key genes and groups of SNPs express themselves gives us the power to eliminate many questions we face in today's burgeoning world of health and wellness information. It provides a SAFE space to land and finds our ideal biological "home." Examining our genetic blueprint allows us to see the signs or guideposts to create a precise health and wellness plan. Depending on what an individual is experiencing, there will be areas that require more support to influence these genetic pathways. What key adjustments can reduce the chance of cardiovascular disease (CVD), diabetes, and rampant stress as I seek to improve my health? How and at what level can we live our truth to support our epic life journey that is deeper than a supplement, a meal, or a movement plan?

For me, it is overcoming challenges. This may have settled my stress neurotransmitters and reinforced the fact that I can do anything. Following the concept of Eustress - those stressors that push our mind and bodies to create strength, I can follow my intuition and do what my physiology gets excited about. Heeding our intuition with the precision support of what our genetic makeup is telling us to do. How cool is that?

Raffaella

Sorrow, deep sorrow, drew me into her pain, love, and loss. I wasn't sure I could support this beautiful woman as her pain seared my heart. How could I help her through this season of her life by explaining what her genetic reports indicate? I questioned my abilities as a clinician and coach when I met Raffaella. Am I too empathetic to lead others? Are my boundaries firm enough to serve others as a vessel without getting pulled into their pain and dilemmas?

I was uncertain for sure. But the signposts in her genetic reports quickly brought me back to the foundation on which to serve and why she sought my guidance. Gotta love science.

We started with sleep. After firsthand experiencing the trauma and devastation of a loved one's death, my dear client was not sleeping.

When I first meet with clients, I inevitably ask how they sleep. Do you wake up refreshed or groggy? Do you maintain a regular sleep schedule? Take medications or beverages to fall asleep? We always start with sleep.

In Matthew Walker's book, "*Why We Sleep*," he reminds us that there is a reason our bodies are meant to sleep for approximately eight hours (most certainly genetically influenced), and to ignore this is playing with health and wellness fire. According to Walker, ascribing to the camp of "I'll sleep when I'm dead" may get you to that endpoint sooner rather than later.

A crucial process that occurs systemically throughout the body during sleep can support total health and well-being. These include activating the glymphatic system, balancing hunger hormones, and reducing inflammation. Dr. Walker explains in his book that the glymphatic system "collects and removes dangerous metabolic compounds" that have collected during wakefulness to essentially put the brain through the top-level car wash option. Dr. Maiken Nedergaard of the University of

Rochester discovered this amazing neural power wash system. He indicates that this system removes neural garbage from our brains during sleep. The brain allows a significant amount of space for the cerebrospinal fluid to come in and "flush out the metabolic debris." Furthermore, this incredible process was found to be most active during the non-REM stage of sleep, highlighting the need for deep sleep, which ideally occurs in the first part of the sleep.

An astonishing fact often repeated is that driving while sleep-deprived is far more dangerous than driving drunk. Dr. Walker notes that our evolutionary survival is threatened during sleep as we cannot hunt, defend ourselves from attack, or move to ideal locations. He ponders the question - how could humans have evolved to sleep instead of existing in wakefulness?

Raffaella slept a few fitful hours upright in a chair, falling asleep to the TV and waking feeling even more depressed than before. After sharing her sleep gene results, I suggested a few small sleep habit changes that could profoundly affect her overall well-being. She responded, "Oh, that is how I used to sleep." So, we returned to her genetic truth and focused mainly on that for a few months. I could sense that she thought this program would be something different. She assumed we would be focusing on calories and a "diet." After a couple of sessions together, she asked me if I would give her a diet based on the number of calories she should eat and if I would expect her to count her calories and track her food. For some folks, this process can be very informative and necessary to understand what they are consuming and why. And this was not what Raffaella needed.

At this point in her healing, Raffaella needed deep, restorative rest. Her hunger hormones were off track, as evidenced by her craving for sweets. Indeed, her genetics indicated this could occur. She was craving sweets because she was sleep deprived. She had potential genetic expression in the following genes MC4R (1) (Melanocortin 4 Receptor), ADIPOQ

(2) (Adiponectin), LEPR (1), and (2) (Leptin Receptor) that when sleep is disrupted, they can influence the hormone, leptin. Lack of restorative sleep can lead to an imbalance in these genes causing a person to feel hungrier and the reduced ability to turn off this genetically controlled impulse. Normally a very active woman throughout her life, she had lost the desire to move her body, to feel strong and healthy again. As her sleep improved with a set bedtime routine, she had less desire for the sweets that, after her trauma, had found their way into her house, and she didn't even want them there. The thought of regular movement and exercise felt exciting again. Her genes whispered a desire for sweets when she was not achieving optimal rest. Raffaella gradually started to feel better as her sleep improved, and you could see a glimmer of hope. Her eyes sparkled again.

It was at this point that I introduced the nourishment of food. I provided a week's worth of recipes for deeply nourishing soups and broths with ingredients such as lemongrass, beets, squash, ginger, and herbs. Raffaella came to me to lose 15 lbs. When her sleep improved, and she nourished her body, the weight began to drift off.

And then the setback arrived. Raffaella visited family, ate less-than-optimal food while traveling, and got off the still tenuous sleep schedule she had begun to establish.

What a gift this was.

I previously mentioned paying attention to the power of words as an essential component of the change process. The state of our minds and the words we use are heard and felt by our minds. Our cells hear the voices in our heads (you know, the ones you wish would shut up at times). Considering the influence of epigenetics, these words can significantly impact our ability to support an optimized expression of our genes. For example, I hear many folks use the personal description, "I am an emotional eater." This can undoubtedly be the case if you have a particular phenotype in the following SNPs MC4R, DRD2(1) (Dopamine Receptor D2), and others.

Thank goodness for Raffaella's setback. It allowed her to see that when she listened to her genetic signposts, she truly felt better - in all ways- and could see what her future could hold. She could indeed move past this emotionally and physically crushing time in her life and thoroughly enjoy her activities and her family.

The gifts in these missteps, mistakes, or failures (if you want to call it that) are crucial in producing lasting change. Without them, we are blind to our full potential, our epigenetic potential. Once her sleep became more regular and nourishing, we could tweak Raffaella's meals and food choices built around what her genes were whispering to us. Should I "be keto," she asked me during one of our sessions. While there can be tangible benefits to following a particular way of eating for some folks, a set of genes indicates what can be optimal for individuals. For example, I have a gene that indicates a need for more complex carbohydrates, especially for weight loss. I have other genes suggesting reducing saturated fat intake to decrease inflammation. However, I can enjoy all kinds of healthy fat in moderation. So, while Raffaella's genetic profile hinted at a higher fat-type diet, my goal was to design and guide her to a plan that she would feel good eating, such as whole foods that were easy to purchase and prepare. I support the design of a healthy and lifestyle eating plan, not a diet in a box with a title. Again, following certain diets can be highly beneficial for some individuals with specific health conditions, so I am not throwing a wet blanket on all types of diets.

For Raffaella, genetically, this appeared to be a plan high in monounsaturated fat, lower in carbohydrates, and a moderate amount of protein. Considering her age, activity, the current level of health, and the foods she enjoyed, we found the perfect plan for her. After working together for about five months, she has released almost all the weight and no longer craves sweets. She still allows herself a treat here and there, which is a wonderful way to live

Raffaella has the energy and clarity to begin pursuing exciting new endeavors and is sleeping much better. She loved discovering the clarity of her genetic blueprint and has referred other clients based on her success.

Dave

Stress comes in many different packages and flavors. There is good stress, or "Eustress," such as when we work out with heavy weights to cause muscle breakdown, which in turn, creates the ability for increased muscle strength. How many songs focus on what doesn't kill you and makes you stronger?

And then there is the pernicious kind of stress, the life-sucking, soul-crushing kind that cruelly lingers, disrupting sleep, increasing inflammatory processes, and negatively influencing genetic pathways that speak to these processes.

Dave came to me during a life-changing event knowing that if he didn't make some profound changes, he would possibly not find the level of health he desired at his age. I immediately connected with him. We discussed his short and long-term health goals and then ran his genetic report. I can still picture his face while I reviewed what his genes told us. Instead of eyes glazing over while I interpreted his results, he wanted to know more about what his genes were telling us. He loved learning this information about himself.

Again, I started with his sleep genetics and hygiene. Dave used an Oura ring, an excellent tracker of different biometrics of sleep, such as time spent in deep, REM, and light sleep. It also tracks his nightly wakeups, physical activity and recovery rate, body temperature, heart rate, respiration rate, and heart rate variability (which can give additional clues to stress levels). As he adjusted what he ate and how he handled his stress, we could monitor whether or not these changes produced a physiological difference. What gets measured gets managed, right? At least, it did in this case.

Dave is a very health-conscious gentleman. Motivated, willing, and cognizant of what living according to what his genetics indicate could mean for his quality of life and health span. Healthy longevity is his goal, so our plan focused on all the shifts he could make to achieve this healthy goal. Still, we were (and still are) dealing with lingering stress due to a major life shift. Dave started using a CGM (continuous glucose monitor) to see how his body reacted to different foods, moods, and activities. Normalizing these numbers is our number one goal. We continue to go back to his genetic reports and dig deeper to find a few more clues and, thus, the appropriate support tools to continue his health optimization. Dave really liked looking under the hood of his genetics! Through the support of a new medical practitioner and my coaching program, he is beginning to stabilize some of the most concerning blood lab values. By no means is this plan a quick overnight fix. I tried the band-aids and the quick fixes and knew they were not the answer. Depending on the complexity of the changes you seek and need to make, this journey is a marathon, not a sprint. The benefits of taking small steps for lasting change and optimized health far outweigh any quick-fix plan.

The Gift of Clarity

Looking under the hood at my own lifestyle, I discovered how genetics has dramatically impacted my health, life, and all my interactions. I found it exhilarating to discover what MY body and mind needed to be fully optimized. My data revealed ways I should dial in on my food, water, air, light, sleep, movement, thoughts, supplements, and even life ambitions. I've gone from a terrifying dark hole of fear and invisibility to waking up nearly every day to my ideal life. How I knew it could be from a soul level. This is not about perfection. It is about taking action and using a highly personalized plan. I respect the evolution of this process, falling off here and there, listening to my body, and making the appropriate adjustments. When I become stubborn or fearful, I dial up my courage and wisdom and remind myself that the gifts on the other side are worth it.

As a young girl, I wanted to be a jockey or a rock star. I'm not sure if they have discovered a "drummer" or "equestrian" gene; if they do, I know I have them. I firmly believe that when you listen to these whispers, along with a highly personalized health and life plan using the wisdom of your genes, you can come much closer to living your soul's true path. My dad was a drummer and a worrier. Thank you, Dad, for all of it.

How Understanding Epigenetics Helped Me [and others] Deal with Menopause in a Powerful Way

by Kym Connolly and April Wright

Courage is discovering that you may not win,
and trying when you can lose.

Tom Krause

Here I was, a dietitian, counseling people on how and what to eat, and I was visibly gaining weight—packing on belly fat. Why? I had absolutely no idea at the time. It was frustrating and professionally embarrassing. I felt undermined by my own body. The worst thing was that I didn't have the answers to fix this. And I'm a fixer! I hadn't changed what I was doing, but I was changing.

For a few years, I'd followed a diet backed by research and known to be helpful in weight loss - the high-fat, low-carbohydrate model similar to the ketogenic diet. And I could see from my own clinical experience that this diet style works quite well for some, but not everyone!

As dietitians, we were traditionally trained to use one diet style (reduced calorie, low fat) for weight reduction for most people, and after more than a decade in practice, I could see this clearly worked for some but not for everyone. Most clients I worked with had already dieted in some fashion and came to me for help because the diets weren't working. It was frustrating to me and demoralizing for clients. Something more was going on. I did not know what.

So, I went looking for answers, and there was little available. That's when I found epigenetics. Little did I know, this discovery would not only impact my work as a dietitian but give me clarity and guidance around menopause, which was something I was also dealing with at the time.

How did epigenetics help the people around me and me? First, epigenetics provided a very practical way to understand and deal with how we can shape our diet to support our bodies and menopause symptoms.

I was Eating Wrong for My Body.

One of the things that many women, including myself, notice when the hormonal changes start around perimenopause (the time before menopause) and during menopause is weight gain. So it's a symptom of this change that hormones are beginning to shift. And once I did my genetic test and got the results back, I noticed that I was not eating in a way that aligned with my body's genetic code.

I had been eating high fat and low carbohydrate, and what did my test say? A better diet style for weight loss was higher carbohydrate and low fat. So, opposite of what I'd been doing and very different from the prominent diet trend at the time. So, changing how I ate really made a difference, and sure, the continual weight gain stopped, but more importantly, I felt better. I had more energy.

Then I turned my attention to understanding how our genes and lifestyle could influence symptoms during menopause and beyond because menopause can hit hard and significantly impact the body.

Our Warrior Wants to Enter Semi-Retirement

During menopause, we women ride a roller-coaster of hormonal changes. They go up, they go down, and everything in between. There are two main female hormones estrogen and progesterone. After the hormonal transition, which is menopause, we have much less of both of these hormones.

This is very significant. Estrogen is our warrior. She goes in to fight battles with our immune system. She's also a peacemaker, helping things to run smoothly so we experience less stress or anxiety. When there's less of this soldier to fight our battles, we feel it.

With menopause, there's less estrogen to go around. She's in semi-retirement. This is where we may notice how well our estrogen receptor genes work in an environment with declining estrogen levels. And those estrogen receptors are everywhere, from the brain to the gut to the mouth!

One of the things you can learn from a genetic test is whether you have strong or weak estrogen receptors. You may have had more premenstrual symptoms (PMS) if you have strong receptors. If you have more fragile receptors, you're less likely to have experienced much in the way of PMS but more likely to have menopause symptoms. And average receptors are sitting between.

Shelly (not her real name) was a client of mine who had poor estrogen receptors and a really painful but (hidden) little-known menopause symptom as a result. When Shelly came to see me, I noticed deep lines on her skin. I thought this was likely connected to the decreased collagen available in the body after menopause. If it was impacting the skin on her face, what more could be going on?

So, I asked her about her vagina. Was she experiencing dryness and pain? And I could immediately see that I had hit a nerve because she started to cry. Many women, including Shelly, suffer in silence with other menopause symptoms like vaginal dryness, painful intercourse, and/or the urge to urinate, which is called GSM (genitourinary syndrome of menopause). There is often a lack of support for these symptoms because doctors don't talk about it, and women are often uncomfortable bringing it up. Shelly hadn't had sex with her husband in over a year, which was quite difficult for them as a couple. But the worst part of it was the pain.

During menopause, the vagina can experience a decrease in blood flow and become paler and dry. It's so painful that sex can become impossible, and some women can't even wear underclothes. But, for some women, it's the dryness that causes issues. This is connected to losing the good microbes, the mucus membrane, and tissue structure. All of which are under the control of estrogen.

So, what was our fix for Shelly? She could increase HLA (Hyaluronic Acid), which helps lock in skin moisture and stimulates collagen production. She could use a cream and eat foods that increased HLA production, like grapefruit, oranges, sweet potatoes, and dark leafy greens. Bone broth, which contains collagen, was another great food to add to her diet. Also, adding fermented foods would help support her vaginal and gut microbiomes. Of course, we also suggested she seek medical support.

What's important to note here is that genes are not your destiny. Poor receptors on a gene test report don't mean a painful vagina or other symptoms of weaker estrogen activity after menopause, such as brain fog, skin changes, an increase in IBS symptoms, or a more vulnerable immune system. But when we know our unique genetic code, we can take action. Using epigenetic knowledge to improve our lifestyle and support our bodies can positively impact our genes or mitigate risks. As a result, we can be proactive in protecting our health.

Menopause and the Common Estrogen Genes that are Associated with Cancer Risk

Tests can sometimes reveal genes that may not work as favorably. This is not genetics. This is epigenetics. As in, we're not just stuck with the result.

By and large, we all have the same genes, but there are several common variations, meaning the genes can behave differently. Epigenetics is the lifestyle choices that influence how our particular pattern of genes is expressed in the body.

One group of genes is the Estrogen Detoxification genes. When it comes to detoxing estrogen, I have a very common, but not as helpful, gene variant. Estrogen breaks down in two parts, and it's important to detox these "middlemen" as quickly as possible because the longer they are in the system, the higher chance they are of producing damage that may lead to abnormal cells.

Since I have this gene variant, I aim to reduce the risk of damaging my DNA by detoxing the byproducts of estrogen breakdown. I want to decrease the creation and production of those unwanted, abnormal cells from developing into something more serious.

I can do a blood test to show me how much of each of these estrogen byproducts (4OH and 2OH estrogen metabolites) I'm making at the time of the test. This allows me real-time knowledge from the actions I've taken. It's vital for me to take this test as I cannot feel or notice changes in my body. The blood test increases awareness. In addition, I can maintain optimal levels by retesting, which reveals how my diet and lifestyle impact those levels.

My favorite supplement for helping my body detox is broccoli sprout powder. Adding cruciferous vegetables like broccoli, cabbage, kale, and cauliflower to our diet is very important to our health. In addition, I make sure I eat plenty of fiber.

By looking at my genetic potential using genomic testing, I can understand the risk I face and have been able to put additional support in place by eating those foods that support my body. In addition, continued blood tests will reveal if I need to add supplements.

April's Trigger-Happy Inflammation Gene

Sometimes you don't notice anything substantial when you make changes based on your genetic test results (but still, you know it's good for you), while other times, you definitely do! April's trigger-happy inflammation gene is a prime example of this.

With menopause, inflammation is something we generally want to reduce and avoid contributing to. Without those protective female hormones, we get more inflammation which we can feel in our joints or as brain fog. Inflammation gets in the way of the body doing its job effectively. It also creates the risk of other, more serious issues like chronic diseases, which have ongoing low-grade inflammation. So, we want to be on top of inflammation.

April is my co-director and one of the first people I did a genetic test for. I wanted her to see the power of this test. While she wasn't looking to solve any major health issues, our discoveries would likely reduce or eliminate significant risks. Wherever and whenever possible, we focus on prevention over treatment.

April has a gene variant that is very sensitive to creating inflammation in the presence of saturated fat. Think beef fat, cheese, butter, and coconut oil. Consuming more than a little bit of saturated fat in a day can create more inflammation and has been connected to risks of plaque in the brain and arteries, which can lead to disease. You may have heard the message of reducing saturated fat to improve heart and brain health.

After a bit of gentle prodding on my part, she started tracking how much she was eating and reduced the saturated fats in her diet. Though the risk is long-term, she began noticing a difference within five days. She realized that, over time, she had become accustomed to the inflammation symptoms. It wasn't until it was gone that she noticed it had been there all along.

Next, we introduced an epigenetic hack for her —high-quality extra virgin olive oil. This type of olive oil has an epigenetic effect. It can turn down these inflammatory genes. Shut them up in a way. We can turn genes on or off by what we do or don't do.

Research shows that about three to four tablespoons of extra virgin olive oil can have a powerful anti-inflammatory effect. So April now uses that as a tool to reduce inflammation. She puts olive oil on anything she can to reach three to four tablespoons in her diet.

April reports decreased puffiness around joints, decreased sinus congestion, more freedom of movement, and more energy—such a great win for a small change.

The Importance of Paying Attention to Nutritional Weaknesses

A genetic test shows how well an individual absorbs certain nutrients or areas they need more nutrient support.

The vitamin choline is a great example. Many people have the gene variant where they need a bit more choline in their diet. Also, in menopause, your body needs additional choline as the activation of choline is decreased without estrogen to help absorb and activate it.

Choline is essential. It helps your brain and nervous system as part of the compound acetylcholine, which is necessary for many functions, especially memory storage. In addition, choline helps your liver get rid of fat and relax muscles (think about cramps or increased blood pressure). It has recently been added to some pregnancy vitamins because of its essential role in brain development.

This issue with choline came across for one of my clients, Mary. We wanted to support her liver. This became a focus when we learned she didn't activate choline efficiently. It's not too difficult to make sure you have enough choline. Mary added a few eggs with runny yolks to her diet weekly. The runny yolk supports choline deficiencies. Had this addition been difficult for Mary, she would have added a choline supplement to her diet to improve her memory and brain function.

Self-Control Is Up to You ... Until It's Not.

Sugar is one of the key areas we focus on with women who are in menopause. This is because women's sugar intake increases post-menopause. Also, body fat storage around the waist increases in response to the insulin hormone, partly due to a direct reduction in estrogen and partly thought to be due to changes in the gut microbes. Estrogen, which helps keep a naturally produced chemical 'uric acid' lower in premenopausal women, is now reduced in menopause, thereby increasing uric acid.

Uric acid, aside from its role in creating gout, also has a role in driving more storage of our food into our fat tissue while also increasing hunger. Simple sugars and exceptionally high levels of fructose drive this mechanism. Increased inflammation in our body can be caused by excess belly fat because fat secretes inflammatory compounds, which we don't want.

What does that have to do with self-control? One of our clients, Suzy, struggled with weight gain around her belly. She was also feeling "puffy" (that's inflammation). Suzy had always looked at herself as a weak-willed person. At least when it came to food, she had always felt bad about her "character flaw." Whenever someone put out anything sweet and delicious, she couldn't resist going back for more...and more. She didn't know she could help it at the time and always felt disappointed in herself afterward.

Her genetic test revealed that she has the "cookie jar" gene. This gene variant is the 'sweet' taste receptor on the tongue and can disable the "self-control" part of the brain once activated by sugar. As a result, it decreases self-control resulting in return trips to that proverbial cookie jar.

This knowledge was crucial for Suzy. It was a biological process in her brain, not a lack of character strength. This allowed Susy to reframe how she felt about herself. She wasn't so weak-willed after all. She then created strategies to prevent this behavior from occurring. One of those strategies was to avoid those sweet foods or only make a small portion of those highly palatable foods available to her.

Sugary foods are enjoyable to eat, but with menopause and the change in hormone levels, we can't eat the way we used to and expect the same result. So for Suzy, who wanted to lose weight, this strategy was important to implement to see the results.

Menopause It's Not Just Stressful; You're More Stressed.

One of the things I noticed about myself while going through menopause is that I would get quite upset and stressed out at things that would have been no big deal in the past. Or at least I would have kept my cool. But I was so easily stressed out! It would set me off if the kids didn't complete their homework or left socks all over the house (again) for their mother to pick up! This was hard because I didn't feel like "me." After the event, I didn't know why I was getting so upset. Was something wrong with me? I really had no guidance about what was happening to me.

I knew I could look up the cortisol hormone in my genetic test. Cortisol is the stress hormone, and I found I needed some support. As I'd always needed this support, what was different now? I did research. I found that with menopause, the stress response can be more significant, meaning I could become more stressed and feel stressors more deeply. Everything was amplified.

Now I knew my genes needed additional support to handle stress and that menopause had increased that requirement. So it was time to introduce some epigenetic hacks in my life to combat this.

One of the things I did, and something we routinely introduce to the women we work with, was a self-care practice. Once we're in menopause, self-care becomes essential to our health care. We have to develop an ongoing practice that we keep in place. This involves calming the system and bringing it down and back to balance. I introduced several self-care practices into my routine. One is meditation. Research shows meditation has powerful epigenetic effects on menopause and benefits sleep duration.

Another practice I put into place was breathwork. I can breathe in a way that amps me up, balances me, or helps me slow right down to sleep. I also created playlists of music I love to lift my spirits and other playlists to calm me in stressful moments.

We can do so much to calm ourselves, but implementing it is key. The result? It gave me a calmer stress response and more peace, which gave me a second to choose my response before taking action. Instead of getting straight into feeling stressed out, my body can now take a beat or a breath to think first.

Muscles Are So Important in Post Menopause.

Liz has always been strong and fit, but once menopause hit, she felt like her muscle was disappearing. Muscle mass does reduce with age, but menopause has the potential to speed things up.

One of the things we looked at for Liz was her genes around the potential to build muscle size and strength and her protein. Once you are over 50, you need to be proactive about protecting the muscle you have. Liz's muscle type does not develop easily. She has small muscle bellies, meaning she has less muscle coming into the changes during menopause. Not to mention muscle protection gets affected by reducing Estrogen. Less muscle means less ability to deal with simple sugars (muscle is where we store our supply of glucose) and can mean faster loss of strength and ability to function with age-related muscle loss. We recommended that Liz work with an exercise physiologist to maintain what she had and continue building strength.

In addition to this, we found that Liz was under-eating protein. Many women do not consume enough protein to get to menopause and beyond. Protein is needed to maintain, repair, and build muscle. According to her genetic test, Liz does well on a higher-protein diet. So making protein a priority became quite beneficial for her. Eating protein throughout the day supported her muscle-building exercises; she had more energy and felt fuller for longer. This meant she wasn't reaching for sugar hits in the afternoon and other little pick-me-ups during the day.

Is Man Flu Real? Can I Get It Too?

One of the things we (fortunately or unfortunately – you decide) discovered in our menopause research is that estrogen plays a role in our immune system. It does make the immune system stronger. So those men who say they feel worse than you when they get sick, that women are stronger, well, there may be something in this. That is until estrogen diminishes in menopause.

We could see this clearly with one of our clients, Shannon (not her real name). She had just gone through menopause and was hit with a virus. And she got hit hard. Then, she got another flu, another cold, one after another. So it was a revolving door of viruses that she didn't fight well.

Several possible issues were going on with Shannon. First, her immune system becomes more pro-inflammatory with less estrogen, as estrogen directs more anti-inflammatory macrophages (part of our immune response). Second, even though she lives in a sunny place, she needs more Vitamin D than the average person, according to her genetic test. And she wasn't getting enough to support her, which also showed up on her regular blood test as a low vitamin D level in her system. Vitamin D is critical for immune function.

While sick, Shannon would take a Vitamin C and a high-level zinc supplement, but this may not have been ideal for her. Why? She's excellent at absorbing and storing zinc in her body. Zinc must be balanced for the immune system to work well, often in large doses, but it may have imbalanced other nutrients necessary for her immune system to function optimally.

We implemented several things for Shannon. We've put together a plan for when she's sick—which supplements and how much. We've also put together a sleep supplement plan, as poor sleep contributes to a lowered immune system function. Now she has a program tailored to her unique body and stage in life.

Not Eating Late - For Your Genes

One of the really cool genes we've learned about is one that influences how food impacts your sleep. Most people don't metabolize food late at night because it's out of our natural circadian rhythm. For some, they have a brake on their ability to metabolize sugar from carbohydrate foods— even those of quality.

One of our genetic test clients, Kat, was really struggling with sleep, and when we went through her diet, it prompted us to go and see if she had the "don't eat before bed" gene. Menopause can impact your sleep enough with the loss of estrogen that we don't want anything else adding to this.

Kat was very busy. With long workdays, she often ate dinner only an hour before bedtime, dinner at 7:30 or 8 PM, and bed at 9 PM. We found that Kat had this "don't eat before bed" gene, which meant she needed a clear strategy to improve sleep—don't eat three hours before bed. Why not? There are sleep hormone (melatonin) receptors in your pancreas. Our pancreas releases the hormone insulin to metabolize the sugars from foods. When melatonin turns on to help initiate sleep, it also says to turn off digestion; we shouldn't be eating now. So that gene variant can impact our ability to release insulin to help dispose of those sugars to a greater extent before bed, keeping our energy up and disrupting our sleep.

While it was hard, Kat adjusted her dinner time. Getting sleep was worth it. She still struggled with quality nights of sleep, yet this change improved the net result. Looking at her genetic test gave us opportunities to see the many factors that can contribute to our symptoms and start to peel back those layers more strategically based on our unique genetic makeup.

Your Bones, Your Genes, and The Sun

A key feature of menopause that can often go overlooked because it doesn't impact how you feel or look daily is your bone density. This is because your bones continually create new bone to replace the older bone. But with menopause, the new bone production slows down, and the turnover speeds up.

Genetic tests can tell you if you're at risk for reduced bone density. All women need to pay attention to this. You may be surprised at what we look at in the genetic test to help deal with the loss of bone and how to protect this as we age because it's not calcium.

A woman, Ivy (not her real name), is 20 years post-menopausal, and her body has just crumpled over to one side. She's lost the ability to reach up high; she can no longer hang clothes on the line. Her bones have collapsed and don't support her well above her mid-back. The estrogen receptors we mentioned earlier play a vital role here. Weaker receptors can impact bone and muscle over a lifetime, especially in a post-menopausal low Estrogen environment. Had the new science been available, Ivy could have supported her bones with Vitamin D and K2 (Vitamin K2 is an over-the-counter dietary supplement used for general heart health and bone health as well as supporting the metabolism of calcium in the body).

Vitamin D and K2 both play a role in protecting your bones. I hear from people all the time that say they are in the sun and should be getting plenty of Vitamin D. Our genes affect how much Vitamin D we need and how well we convert sunlight into Vitamin D. Ivy lives in a sunny and warm part of the world, but that doesn't guarantee she's getting enough Vitamin D. You may need more sunshine to really absorb Vitamin D and more than someone else does to have enough to support your bones.

Vitamin D's great partner is K2. They work well together. And there are genetic variants that mean you may need more Vitamin K2 as well. You can measure Vitamin D in the blood, but there is no current measure for K2.

For Ivy, had we been able to help her in the past and known through her genetics what was needed, we could have recommended a Vitamin K2 supplement to act as the glue that sticks calcium to the bone and Vitamin D to activate bone receptors to take up calcium. Unfortunately, it's difficult to have a diet rich in K2 as most people don't eat enough plant matter (like spinach and leafy greens), liver, or fermented foods like natto, and we can't measure it accurately. In this case, a supplement would have been recommended as insurance.

It is incredible that, based on our genetics, we can put these preventative actions in place, so we have the potential to live healthier for longer and suffer less from debilitating bone loss.

Menopause is a Time to Put Longevity Actions in Place.

With all the research I did in my search for answers on dealing with the symptoms of menopause, one thing that really surprised me was glycan age testing (It determines your biological age by measuring chronic inflammation in your system), which showed that menopause can speed up our biological age.

Chronological age is how many birthdays you have had. Biological age is how old your cells are, which might be older or younger than your birthday year. The faster our cells age, the quicker we run into the primary diseases of aging - heart disease, cancer, dementia, and diabetes. Keeping our cells more youthful protects us from chronic disease for a longer period within our lifespan - it extends our health span.

The hormonal changes in your body remove you from the protective estrogen bubble you've been living in, and now you have to deal with a whole host of things that can make you look, feel and be older. So even for women who barely feel the change menopause brings, there is still an acceleration in biological aging.

Because of this information, I have used my genetic test to optimize my nutrition and other areas to help slow aging. For example, I know that I can speed up my biological age by making poor food choices (think poor quality fast food), exposure to smoke, or drinking too much alcohol. But, armed with my genetic knowledge, I also have the potential to slow it down.

I've done this by eating right for my body, not what someone else says is a good diet plan. I supplement with what my body needs for extra support. I exercise to promote muscle maintenance and strength over the long term, not for a beach body. I practice self-care to reduce stress and inflammation in my body. I practice periods of fasting-mimicking interventions. That is, eating a nutritionally specific, low-calorie diet mimics fasting and tricks the body into turning on the survival pathways, cleans up the garbage in cells, and they are healthier. All the benefits of fasting, which increase health span, with a little food, so it's less challenging to do. By doing this, I can influence beneficial genes to turn on and help my body cope with these changes. I know I can also turn more negative genes down or off through what I do. This is the power of epigenetics.

The Many Genes of Histamine Intolerance Genomics was a Game Changer for Me

by Eileen Schutte, MS, CN, FMN

Don't believe every worried thought you have.
Worried thoughts are notoriously inaccurate.

Renee Jain

Shortly after I started my nutrition consulting practice, while attending a conference, I was introduced to a new form of genetic testing called genomics that went beyond testing for genetic mutations like sickle cell. At the time, I did not see the enormous potential of the science of genomics in developing truly personalized nutrition and lifestyle recommendations. Nor did I see how genomics could finally answer why one person might struggle with autoimmune conditions while others struggle with metabolic disorders like obesity.

My genomic journey started with a simple genetic test, which only tested two genes, MTHFR (methylenetetrahydrofolate reductase) and COMT (catechol-O-methyltransferase). I discovered that I had the rarest variant in both genes! And it began to explain why I had an increased need for Vitamins B, B2, B6, B12, and folate.

When I was 23 years old, I suffered a severe staph infection that left me with partial paralysis of the left side of my face or Bell's palsy and tinnitus (ringing in the ears). At that time, my doctor advised me to take B Vitamins to help me recover from Bell's palsy and reduce tinnitus.

Later I developed a common autoimmune condition known as oral lichen planus, which causes sore throat and painful mouth sores. Because I also had chronic canker sores, I started taking high dosages of Vitamin B2 (riboflavin) to heal the sores more quickly. Later, I discovered that Vitamin B2 also helped with my lichen planus outbreaks. As a result, I also increased Vitamin B foods in my diet. I was a vegetarian then and soon learned I needed to bring animal products back into my diet as they are the only food source of Vitamin B12 that was truly bioavailable.

Both gene variances in MTHFR and COMT that I expressed increased my need for B vitamins. Gene variances in MTHFR significantly increased my need for folate, Vitamin B2, and B12 due to their role in a process known as methylation. The COMT gene, on the other hand, depends on healthy methylation to function optimally. You can see how these two genes are intertwined.

My first genomics test ignited my desire to learn more about my genes to explain not just my increased need for B vitamins. For example, could it explain why I struggled with other symptoms like migraines, hormonal imbalances, anxiety, and irritable bowel syndrome (IBS)? And why could I not manage and handle stressful situations very well?

Migraines: Could it be What I was Eating?

Over the years, I struggled with poor digestion, IBS (irritable bowel syndrome), food sensitivities, and debilitating migraines. So then, I began to think, were my migraines caused by what I ate?

Research I reviewed showed a link between chronic migraines and Vitamin B deficiencies. While I had improved my diet, I continued to supplement with B vitamins; I still had migraines. I had thought it might be due to hormonal imbalances and my cycle, yet there wasn't a pattern.

I began journaling my food intake, looking for foods or a combination of foods that might be causing my migraines and digestive issues.

Whereas it took some time, my aha moment came when I realized that I often got headaches or migraines shortly after eating cheese and pepperoni pizza with a glass of red wine. Another time I was at a health fair where I tried kombucha (fermented tea), aged cheese, fermented vegetables, and organic, grass-fed summer sausage and ended up with a raging migraine the next day. What could be the connection between all these foods?

I worked with clients struggling with food sensitivities and intolerances during this time. I began to see a pattern with the same foods that might be causing my migraines and IBS. Most of the foods that appeared to cause their symptoms and mine were high in histamine and tyramine. Research shows that foods high in tyramine, like aged cheese and red wine, are among the leading dietary causes of migraines and headaches. Now we know that foods high in histamine, a toxic compound in food, also cause migraines and headaches for some people.

But why would foods high in histamine and tyramine cause these symptoms? Most foods high in histamine, like fermented vegetables, preserved meats, and fermented dairy products, have been eaten for centuries. Before refrigeration, we preserved foods with fermentation, salt, or smoking, i.e., smoked salmon. These food preservation methods increase the histamine content of these foods.

Along with foods high in histamine, the body releases histamine predominately as an immune response to allergies and infections. Histamine is also released as an excitatory neurotransmitter, a brain-signaling chemical. As a neurotransmitter, histamine plays a role in our sleep/wake cycle and stress response.

Typically, our bodies manage the histamine level by degrading histamine that is released or comes from our food. If we cannot break down histamine, levels can reach a toxicity point causing symptoms like migraines, headaches, skin rashes, and allergies. Even our digestion and gut can be impacted by high histamine levels causing heartburn, gas, bloat, diarrhea, and constipation.

What is Histamine Intolerance?

Histamine was first discovered when people became ill from consuming fish in the scombroid family, which included tuna and mackerel. When fish, especially those in the scombroid family, is not refrigerated and safely handled, this can lead to bacteria converting the protein building block known as histidine in the fish to histamine, increasing the histamine content to a toxic level.

Food intolerances like histamine and lactose intolerance are often confused with food sensitivities and allergies. Food sensitivities and allergies are caused by an immune response to specific proteins and sugars found in food. Food intolerances, on the other hand, are often due to a lack of enzymes that helps to degrade and break down food chemicals or compounds like histamine, tyramine, and lactose (a sugar found in dairy products like milk).

The two main enzymes that break down histamine are diamine oxidase (DAO) and histamine N-methyltransferase (HNMT). DAO enzyme predominately degrades and blocks histamine in the intestinal lining or gut from the foods we ingest. On the other hand, histamine released from our body is broken down predominately by the HNMT enzyme.

Becoming intolerant to foods high in histamine, like fermented vegetables, is caused by a reduced ability to manage histamine levels due to a lack of degradation and elimination of excessive histamine. This is similar to being unable to digest foods high in lactose, like ice cream, due to a lack of lactase, the enzyme that breaks down lactose.

The healthy production of these two enzymes that degrade histamine depends on many factors, including nutrients, gut health, and gene variances that can reduce their function. For instance, nutrients like Vitamin

B6 act as co-factor in the production of DAO. On the other hand, HNMT is highly dependent on all of the B vitamins. Those B vitamins keep showing up!

With more comprehensive nutrigenomics testing, I soon learned that not only did I have an increased need for B vitamins but other nutrients like zinc, Vitamins C and A. These nutrients play a critical role in managing histamine levels. In addition, I learned I had genetic variances in both enzymes – DAO and HNMT, impacting their ability to break down histamine.

Could Stress Make Me More Histamine Intolerant?

The more I learned about my genes and genomics, the more I realized that I was stress-sensitive or could not manage stress levels very well. Due to my significantly reduced COMT gene, I finally understood why I tended to be anxious, experienced a lack of focus, and why I was stress sensitive. The COMT gene helps regulate neurotransmitters like dopamine by breaking them down and ensuring that levels are not too high. Dopamine is a "feel good" neurotransmitter that allows us to stay focused, plays a role in our digestion, and helps to manage pain. A significantly reduced COMT gene meant I would likely have high dopamine levels. However, too much dopamine can lead to increased anxiety, lack of focus, and reduced ability to handle stressful events.

When I learned more about the COMT gene, I finally realized why I experienced anxiety more than the average person. I experienced debilitating anxiety even with simple life events like getting to the airport in time to board my flight.

Histamine, as a neurotransmitter, is released when we are under stress and experience anxiety. The one stressful event in my life that best illustrates how stress can increase histamine release leading to histamine intolerance, was when my husband and I lost our home in a fire.

Losing our home and its contents was stressful, but rebuilding it significantly increased my anxiety and stress levels. I had chronic migraines, insomnia, and a psoriasis outbreak. Increased histamine levels often trigger autoimmune flare-ups, especially psoriasis (an autoimmune skin condition).

I didn't know about my genes then, but when I look back, I see that I was indeed histamine intolerant at that time. Knowing I had a reduced COMT helped me recognize when my dopamine levels were too high. I recognized when to limit coffee and other dopamine-increasing foods or lifestyle habits like alcohol. By reducing dopamine levels, I lowered histamine levels, reduced anxiety, and was less stress sensitive.

I also learned simple ways to support my COMT gene to reduce anxiety and increase focus. It sounds crazy, but reducing clutter and keeping an organized desk and environment significantly reduced my stress levels. Over the years, I found getting rid of clutter almost cathartic, giving me such relief. Now I know why.

While coaching, I began recognizing the same symptoms in some of my clients. For example, my client, Jackie, had similar genes to mine and a significantly reduced COMT gene. When I started working with her, she struggled with anxiety, hormonal imbalances, and digestive issues and was definitely stress sensitive. Addressing her COMT gene through simple changes like reducing coffee and foods high in sugar helped her to handle stress better and reduce her symptoms. Working with her, we soon recognized that she was also histamine intolerant, so by lowering histamine-rich foods, she reduced her anxiety and allergic-type symptoms and improved her digestion.

Hormones and Histamine Intolerance

Not only is the COMT gene instrumental in managing our dopamine levels, but this gene also degrades estrogen. Therefore, during stressful times or if dopamine levels are high, dopamine usually wins, and estrogen does not get degraded. However, when estrogen is not broken down, it can lead to estrogen dominance or hormonal imbalance, especially if you have other gene variants in estrogen metabolism and reduced estrogen elimination.

Guess who was also estrogen dominant? When it comes to histamine, estrogen plays a unique role. During pregnancy, estrogen will help to increase the DAO enzyme to degrade histamine. But it's not the same when you are not pregnant estrogen reduces the DAO enzyme. This means the ebb and flow of estrogen levels can impact histamine levels. It is not uncommon for histamine intolerant women to be more sensitive to foods high in histamine during certain times during their cycles. When I had my cycle, I noticed I was more histamine intolerant during the mid-cycle when estrogen was at its peak. Lowering my intake of histamine-rich foods during this time reduced my symptoms, especially migraines, and headaches.

Exercise and Histamine Intolerance

My father was instrumental in introducing me to the many sports and forms of exercise I love today. As a child, I loved swimming, hiking, and bicycling, but when I found tennis, that quickly became my favorite sport. Whereas I did not start until later in life playing tennis, I quickly became addicted and ended up competing in competitive tennis leagues. And that meant I played a lot, up to four to six hours daily.

When we exercise in competitive sports where stress is high, our bodies release histamine. Remember, histamine is an excitatory neurotransmitter, so along with adrenalin. We release histamine to keep us going in sports or activities and to keep our minds sharp and focused, ready to return those super-fast tennis serves. That means we are more likely to be intolerant to foods high in histamine after playing a sport or exercising. Consuming histamine-rich foods after exercising can easily lead to symptoms of histamine intolerance, especially if you enjoyed a couple of slices of pizza and a glass of wine.

As you can probably tell, I have enjoyed a glass of wine and splurged on pizza a few times after a tennis match. But I paid for it that evening or the next day with a migraine, brain fog, and fatigue as high histamine disrupts sleep. Knowing that I was histamine intolerant and how my genes came into play has helped to avoid that "histamine hangover" after a tennis match or endurance exercise.

Can Your Genes Impact Your Digestion and Gut Health?

Over the years, I struggled with poor digestion and gut health. At the time, I attributed it to my hormone imbalances and diet, which sometimes lacked healthy fiber. However, estrogen dominance can play havoc with digestion, leading to chronic constipation, diarrhea, or what is referred to as IBS-Mixed.

Because I had taken long-term antibiotics numerous times, I also attributed my poor digestion to possible bacterial imbalances in the gut microbiota due to antibiotics. Antibiotics are excellent at killing pathogenic (harmful) bacteria but can also destroy commensal (good) bacteria, leading to gut microbiota (bacteria) imbalances. Research shows that microbiome imbalances in the gut can lead to poor digestion, IBS, and potentially intestinal permeability or "leaky gut." Intestinal permeability or leaky gut almost always leads to food intolerances, including gluten, which is different from Celiac disease, a wheat allergy. In most cases, healing a leaky gut will increase tolerance to wheat, allowing one to go back to eating foods that contain gluten.

I completed microbiome and food intolerance testing to see if I had a leaky gut, gut bacteria imbalances, and possible gluten intolerance. My lab tests showed that I was gluten intolerant, had gut microbiota imbalances, and had a "leaky gut." These test results explained why I was experiencing poor digestion and possibly one of the reasons I had acquired autoimmune conditions like psoriasis and lichen planus.

During this time, I also discovered that the microbiome or gut microbiota hugely influenced whether you could become histamine intolerant. Microbiome imbalances and bacterial overgrowth can promote a leaky gut and reduce the formation of the DAO enzyme in the gut lining. Remember, DAO is the critical enzyme that breaks down histamine from our food. In addition, specific probiotic strains, bacterial overgrowth, and some infections like yeast overgrowth produce histamine, leading to higher levels of histamine in the gut.

After additional microbiome testing of my gut, a pattern began to emerge. Despite dietary changes and probiotic supplements, certain bacteria were always out of balance. Could some genes influence the makeup of my gut microbiota? My tendency to be gluten intolerant?

At first, I really couldn't comprehend how a gene might influence my gut microbiota. For example, could genes, including those that manage histamine levels, impact your digestion?

When my genomics test revealed that I had a variant in the FUT2 (fucosyltransferase 2) gene and that I was a "non-secretor," I realized that there are genes that can impact the gut microbiota. The FUT2 gene is a unique gene that helps provide food or a compound known as oligosaccharide to our gut flora. If you are a non-secretor, your gut flora will not flourish as well due to lower levels of oligosaccharides, leading to commensal (good) bacteria imbalances. One of the imbalances is a reduction in bifidobacteria. This critical beneficial microbe lines the gut helping to increase nutrient absorption, providing us with nutrients like Vitamin B2, and lowering the risk of acquiring a leaky gut. Having a variant in the FUT2 also made it more likely for me to be gluten intolerant, have digestive issues, and struggle with maintaining a healthy gut microbiota. Also, they are more likely to have histamine intolerance.

Did My Gut Health Make Me More Likely to be Histamine Intolerant?

Histamine intolerance is uniquely different from other food intolerances like lactose intolerance, mainly because we not only get histamine from our food, our bodies release histamine. So, you could go on a low histamine diet and potentially still experience histamine intolerance symptoms due to a high release of histamine in your body.

Remember, histamine intolerance results from our body's inability to manage histamine levels from our food and what our bodies release. Histamine is a compound that plays many beneficial roles, including a critical role in our immune system. But when histamine levels in our body get too high, we can experience toxic symptoms, including headaches, digestive issues, skin rashes, and allergic-type symptoms. Essentially our histamine bucket overfills. Whereas histamine, as a neurotransmitter, plays a major role in our mental health, its most significant role in our body is to help support our immune system.

The microbiome, or the bacteria that is a part of our gut lining, plays a critical role in protecting us from toxins and other compounds from the foods that we ingest. Eating is probably the most harmful yet dangerous activity we do every day as we expose the outside world to our gut. The microbiome protects us from the toxic compounds we ingest and the toxins produced during digestion. The microbiome also helps to prevent a leaky gut and reduce harmful compounds from being released into the body. Essentially our gut microbiota plays a significant role in our immune system, helping to protect us.

But suppose our gut microbiota becomes imbalanced through the lack and diversity of bacteria or an overgrowth of harmful bacteria. In that case, this can lead to a leaky gut and heightened immune response. Part of that immune response is a high release of histamine. Like an allergic reaction, where our bodies see environmental or certain foods as foreign, histamine is released to protect us and mount an immune response.

I also struggled with histamine intolerance symptoms when I discovered I had leaky gut and gluten intolerance. My poor gut health made me more intolerant to histamine-rich foods as I released high amounts of histamine with a decreased DAO production. Top that off with my gene variant in HNMT; my histamine bucket was overfilling.

Because I also had genes that increased the likelihood that I could have gluten intolerance, I eliminated gluten from my diet—over time. Whereas I have gone gluten-free numerous times in my life, I have always thought that one should get down to the cause of food intolerances and not go down the "restrictive diet" path, as this can lead to nutrient imbalances and deficiencies.

Yet, after researching the other genes that increased the likelihood of gluten intolerance, Going gluten-free helped with my digestion and reduced psoriasis outbreaks. I have also become more tolerant of histamine-rich foods like aged cheese, white wine, and yogurt.

Understanding the impact of gut health and histamine intolerance made me realize the importance of reducing the intake of foods rich in histamine and looking at what might be causing the increased histamine release. After all, there is no such thing as a histamine-free diet. Almost all foods contain histamine; even avocados, eggs, and citrus fruits are high in histamine or are a histamine liberator. Histamine liberators are food like bananas and tomatoes that increases the release of histamine within the body.

Can Your Gut Health Impact Nutrient Absorption?

When we have poor digestion and imbalances in our gut microbiota, histamine release increases to protect us and promote a healthy immune response. But it also means reduced production of the DAO enzyme. Research shows that high histamine in the gut almost always comes with lower DAO production.

Imbalances in the gut flora can also impact the production and absorption of nutrients. Some of these nutrients include nutrients that support the production of DAO, like Vitamins A and C. Commensal bacteria also produce most B vitamins in the gut, such as Vitamin B12, folate, niacin (B3), pyridoxine (B6), riboflavin (B2), and thiamine (B1) in the gut microbiota.

Over the years, I experienced one of the symptoms of Vitamin B2 deficiency: painful cracking on the corner of the mouth or angular cheilitis. Why would I have a deficiency in Vitamin B2 when my diet was ample in food sources of riboflavin like salmon, eggs, and other animal products? The answer may be in my gut and my genes.

The more I learned about the microbiome, digestive health, gluten intolerance, and my genes, the more I realized that histamine intolerance originates in the gut. Since I had the FUT2 gene, which increased the likelihood of gut microbiota imbalances and becoming gluten intolerant, the key to resolving histamine intolerance was to address my gut health.

Not only was my poor gut health increasing the release of histamine, but it was also impacting my body's ability to produce the key enzymes that degrade histamine, DAO, and HNMT. Remember that the production of both DAO and HNMT enzymes depends on key nutrients or cofactors like Vitamin B2, B6, C, and A.

Can Knowing Your Genetics Help Heal the Gut?

The best way to rebuild the gut microbiota and heal the gut is to increase your consumption of fermented foods like yogurt, sauerkraut, and fermented soy products like tempeh because they contain healthy fiber and probiotic strains that help rebuild the microbiome. I knew that fermented vegetables like sauerkraut were out for me as I tended to be more sensitive to them. So, I thought I would try making yogurt with a bacteria strain known as *Lactobacillus bulgaricus*. Little did I know I made my yogurt with histamine-producing probiotics! Using that probiotic strain didn't work out too well.

Because fermented foods contain probiotics that help the gut, it can be challenging to heal without them. But it can be done. Knowing I had the FUT2 gene made me realize that going gluten-free was a significant first step in calming the immune system down and healing the gut. I also supplemented with B vitamins, Vitamin C, and zinc, supporting my gene variants in both enzymes, DAO and HNMT.

But the most critical step in healing the gut is improving digestion, whether chronic constipation, gas, bloat, or diarrhea. Unfortunately, poor digestion almost always leads to poor nutrient absorption and poorly digested foods that provide ample food for bacteria in the small intestine to feast on and thrive, leading to what is known as small intestinal bacterial overgrowth (SIBO).

My next step in healing my gut was to address my digestion with foods rich in fiber and nutrients like folate, including increasing my consumption of cruciferous vegetables like broccoli, cauliflower, cabbage, and kale. Cruciferous vegetables are high in folate and high in super nutrients called bioactives. These bioactives help support my gene variants in MTHFR and the genes variants that impacted my production of the powerful antioxidant known as glutathione.

I also needed to address and support my FUT2 gene by restoring diversity and abundance in my gut microbiota. Since I couldn't tolerate fermented foods, I chose the next best thing: foods high in prebiotics or that feed probiotic strains. Supplemental probiotics can help rebuild the microbiome also, but perhaps not as effective if there is no food for them to consume. Prebiotics are more effective at increasing and rebuilding the gut microbiota. After all, who wants to stick around if there isn't enough to eat?

I increased foods high in prebiotics and fiber, like legumes, nuts, seeds, oats, and apples. Interestingly, new research shows that dietary fiber can help resolve histamine intolerance and support our immune system. I also supplemented with specific probiotic strains that degrade histamine while ensuring I was not taking probiotic strains considered to be histamine-forming, like *Lactobacillus bulgaricus*.

Over time my digestion improved, and I began tolerating more histamine-rich foods like tuna fish and preserved meats. And I also had a dramatic reduction in my symptoms, especially headaches, migraines, skin rashes, and insomnia. But I couldn't have gotten to this point without knowing my genomics, as it gave me the road map to overcoming histamine intolerance!

Does Genetic Testing Determine if You are Histamine Intolerant?

I would love to say that by completing a genetic test, you can determine if you are intolerant to foods high in histamine. But as you can see from my journey, many factors can increase the likelihood of becoming histamine intolerant and may or may not include genetic variances. In addition, genetic testing does not usually address if you have severe disease or health conditions unless you are testing for a specific genetic mutation like Huntington's disease. This rare, inherited disease causes the progressive breakdown (degeneration) of nerve cells in the brain. Therefore, your doctor or other healthcare professionals should test for severe genetic mutations.

Because histamine intolerance symptoms differ for everyone, it can be more challenging to determine if you have histamine intolerance based on symptoms alone. The most common symptom is environmental and food allergies; however, not everyone with histamine intolerance will have allergies. Ironically, I have never had allergies, for which I am grateful. Like myself, others will experience other symptoms like headaches, migraines, and anxiety without having environmental and food allergies.

Currently, there are no definitive tests for histamine intolerance. Recent testing for serum (blood) levels of DAO shows some promise for determining if you have histamine intolerance. However, only some people with histamine intolerance will have low serum DAO. As I did years ago, one of the best ways to determine if you have histamine intolerance is to complete a symptom/food journal.

Along with a symptom/food journal, you must consider if you have health conditions that can make it more likely for you to acquire histamine intolerance. For instance, it has been shown that patients with autoimmune IBD (Crohn's disease and ulcerative colitis) showed improvement in their symptoms while on a low histamine diet. Other autoimmune conditions like psoriasis will increase the likelihood of being intolerant to foods high in histamine. In addition, chronic allergies, asthma, and IBS will also increase the possibility of becoming intolerant to foods high in histamine.

Certain medications can either increase histamine release or can inhibit the DAO enzyme. For instance, some antihypertensives (blood pressure lowering) can inhibit DAO enzymes, making you more likely to acquire histamine intolerance, especially if you have other gastrointestinal disorders like IBD or IBS. Histamine intolerance symptoms could be a result of taking long-term maintenance medications. You could use an alternative medicine that does not increase histamine or inhibit DAO by speaking with your doctor.

Genetic testing was a game changer because it led me down the right pathway to determine if I had histamine intolerance and how to overcome it. Taking steps to rule out other possible causes, along with a symptom/food journal, is the best way to determine if you are intolerant to foods high in histamine. Genetic or genomics testing will give you the DNA blueprint or road map to guide you to the root cause and help you overcome histamine intolerance.

Nutrigenomics: Era of Personalized Nutrition

As you can see, my journey in resolving and overcoming histamine intolerance included many different twists and turns. Not to mention quite a bit of guessing on what will work to reduce my symptoms and optimize my health.

Histamine plays many different roles in our body, and managing histamine is truly complex, making it one of the most challenging food intolerances to overcome and resolve. Unlike other food intolerances like lactose which only has one gene that plays a role in the development of lactose intolerance, histamine intolerance involves many different genes.

I would love to say that addressing gene variances in DAO and HNMT, which play a role in histamine degradation, is the answer to overcoming histamine intolerance. Unfortunately, that would be too simple; however, I have seen this often recommended in some blogs and articles on histamine intolerance. Or just addressing variants in MTHFR as I did initially and before I had completed more comprehensive genetic testing.

But the more I learned about my genetics, the more I realized that nutrigenomics, The study of how food affects a person's genes and how a person's genes affect the way the body responds to food, was the key to understanding why I became intolerant to foods rich in histamine. Why some of my clients also struggled with food intolerances, including histamine, tyramine, and glutamate.

Before nutrigenomics testing, I would use food sensitivity testing to help my clients reduce their food sensitivities and intolerance symptoms. However, whereas I could give my clients some relief, they didn't answer why my clients had food intolerances and sensitivities. When I had my clients start with a food sensitivity test, they often would not show sensitivities to wheat and other grains containing gluten. Yet most had variants in the FUT2 gene, making them more likely to be gluten intolerant.

In addition, clients would not show sensitivities or intolerances to foods high in histamine, yet they were struggling with histamine intolerance based on their symptoms. In most cases, their genes revealed the true story that they were more likely to be intolerant to foods high in histamine and sometimes tyramine. I soon learned that food sensitivities and intolerances are often rooted in our genes, and by addressing those genes, I had more success helping my clients reduce their symptoms.

My journey made me realize the enormous potential of nutrigenomics and how it can provide a truly personalized approach for clients to experience optimal health. To help clients with alternatives to chronic health conditions like autoimmunity, where conventional medicine has few answers except one more medication.

How a Biophysics Perspective on Epigenetics Can Change Your Entire Life

by Tova Sardot, Ph.D.

Love manifests form to optimize every facet of your being!

Tova Sardot, Ph.D.

Art and Context

My goal is to be a heart-centered guide, supporting clients by optimizing their human experience with an epigenetic lens on every aspect of their life. I will share how I work at the intersection of consciousness and epigenetics through science, theory, and observation. There is an art in how each brain pulls information together to shape our realities and our experience. Although we all do this, many are unaware that we create our experiences. Everything I present here argues for this key concept—that we can significantly improve our life experience by growing a deeper understanding of the reality we create. This is accomplished by becoming aware of the fundamental components affecting our experience and evolution. There is no set depth or end to this unfolding. This makes understanding ourselves and our potential in the world so exciting!

The process I guide others through brings them closer and closer to Love. Most people think of love in a romantic or familial context. Although, the capital "L" in Love denotes the expression of divinity within you and the truth of who you are at the core of your being. Underneath all the learned ways of existing, this beingness is there and always present. This is why much of my work is focused on freeing you from the acquired distortions that limit access to your beingness. You can approach it by feeling Love for this wonderful human experience, each other, and ourselves. A separation pulls us away from our beingness and these feelings. Connection back to ourselves will bring us back to wholeness. This is what I mean by "Love manifests form." If we create our experience from an awareness that leads us to deeper Love, we cannot help but improve our lives on every level. This is a brief introduction to my work's spiritual/consciousness basis, which gets integrated with scientific understanding.

The disciplines of biology, chemistry, and physics are so expansive that they are often taught separately from grade school to the university level. However, when the sciences are studied together, one may better approach the truth of our existence. For example, the study of human electromagnetic fields would be better understood from an eclectic approach. At this point, the science of biology does not consider how these fields occur and interact daily with living creatures. Physics explains how foundational they are to reality; thus, approaching this subject with multiple lenses provides a more complete analysis.

Physics dictates that any electric current produces a magnetic field, and in the study of biology, electricity is the language of the nervous system. Elements like magnesium, calcium, and potassium each have a charge and flow in and out of millions of membranes to generate electric currents. This, in turn, creates an electromagnetic biologic field that senses other fields and the general environment around us. While most people do not consciously notice these subtle interactions, our bodies do.

Dr. Becker was an osteopathic surgeon in the 1990s who researched how the body generates new tissue and healing via magnetic fields. I highly recommend his book, The Body Electric, to understand humans' electromagnetic components and how they contribute to physical healing.

As we will see with Nicole's case a little later, magnetic fields can also be applied to support a healthy inflammatory and pain response. For some, using these unseen fields for healing can seem like magic. Although, to me, magic is simply science we do not yet understand.

The limitations of the five "normal" human senses lead us to form incomplete perceptions. Our senses are simulations that our brain uses to construct our experienced reality and help us to move about the world. These senses have evolved to keep us alive, although these perceptions form only a partial observation of existence. The only science that approaches using less sensory observation is physics (especially quantum physics), which uses mathematics to derive patterns that describe the basis for reality.

Here I will apply principles from physics to biology and chemistry while reinforcing our understanding of the human body and epigenetics. I aim to introduce the general concepts that inform my perception of the road to well-being. There is, of course, much more depth to the science than what is presented here.

Epigenetics

DNA is deoxyribonucleic acid, a molecule that contains the genetic code unique to every individual. It is an instruction manual for making all the proteins that form our bodies, allowing us to thrive. The ubiquitous nature of DNA across every species is astounding. Humans and chimpanzees share 98.8 percent of the same DNA; cats and cows are 90 percent and 80 percent similar, respectively. DNA is essentially the "hardware" that establishes how cells communicate and respond to the body's inner and outer environments. The genome is the starting place of what could be, but every aspect of an organism's environment determines what it becomes. In this process, epigenetics modifies a being's physiology and is the origin of every evolution. While many DNA changes happen over generations, the initiation for future evolution is shaped by life as you have lived it and your present experiences. While you read this, you are making epigenetic changes to the transcription of your DNA—though many are being made in pencil, meaning they are generally subject to change. There are some marks in pen; these are more difficult to alter and have likely been passed to you from previous generations. I believe this knowledge and your innate power to shape your future epigenetic self can motivate you to make optimal life changes.

Epigenetic research has shown that everything you think, feel, and consume, impacts your genes through the regulation of your DNA, which modifies your cellular structure. The comprehensive nature of epigenetic feedback determines our reactions to disease and even who we are and what we will become; when the ideal inputs are given to the human system on multiple levels, possibilities open to having optimal regulation of the body. This is energy-transforming matter, and we will look to the language of physics for a better understanding of this interaction.

Epigenetic expression, at its core, is determined by the movement of energy in cells. Energy is first dedicated to the functions most needed for survival, and less essential physiologic systems are fed last. This prioritizes survival. However, energy excess or deficits anywhere can lead to epigenetic 'marks' within a DNA strand and irregular transcription and imbalances within the body. Thus, homeostatic balance is essential for a healthy organism, which must efficiently move energy received from light and water through electromagnetic fields. These energy exchanges are made possible by the mitochondria within cells that work together to support an organism. Mitochondria is an organelle found in large numbers in most cells, in which these biochemical processes occur.

How Physics Directs DNA: Light, Water, and Electromagnetic Fields

Light is an energy source for the body. We acquire light either by consuming food or absorbing it via the sun's rays, and these are the paramount factors in loading the human body with energy for optimal function.

The value of food comes not only in the form of minerals and vitamins but also in how we consume sun-packaged electrons from plants and animals. For example, when we eat vegetables or animal meat, electrons are delivered to our inner mitochondrial membranes. They are then made into ATP (adenosine triphosphate), the foundation of energy in both cells and the entire body.

Light is also a direct energy source for the body. Skin absorbs sunlight, which breaks bonds in fat-derived cholesterol metabolites and produces Vitamin D. Light enters the eyes and quickly hits the SCN (suprachiasmatic nucleus) to regulate our circadian rhythms. This action causes the hormone melatonin to adjust our sleep cycles based on the time of day and year.

The human biofield refers to the body's emissions of a low-level field of light or photons, which Dr. Fritz-Albert Popp first observed. He concluded that living organisms create their own light as an ultrafast communication system at the basis of all regulation in the body. These communications are how all regulation occurs within and between systems of the body. Chemical reactions create signaling in seconds, but light communication occurs thirty million times per second. Dr. Popp also found that DNA both absorbs and emits light. Throughout these processes, each cell in the body vibrates, shares information, and contributes to the biofield, and your epigenome listens and responds.

Coherence measures the level of homeostasis in the body through light-based information sharing. When all organs receive optimal energy levels, no part is overworking to compensate for an imbalance, and there is proper signaling within and between all the body's systems. Large amounts of information can be carried and transmitted if there is coherence. Information sharing ensures that each part of the body, down to the cell, knows what is happening and has instructions and energy to perform those instructions. When it fails, however, the body's systems begin to fall apart, which indicates danger as a complete loss of coherence results in death for any organism. The human biofield can be observed using ultra-sensitive cameras, which can provide information about how well a body works based on whether it has a balance, deficiency, or overabundance of light. The areas lacking or having too much light can also indicate which organs are struggling and where coherence is weak.

According to research by the Heart Math Institute, a lack of mental and emotional management is a cause of loss of coherence in the body. When the mental and emotional energy is not managed, there is a severe drain on our nervous system's natural, vital energy. This creates deranged neural and hormone feedback in the brain, reinforcing negative feelings. Further, the person may experience a loss in clear and rational thinking, self-regulation, and intuition. Many live through decades of lowered coherence, and disease states can manifest. This is why gratitude and appreciation practices are becoming popular. When you engage positive thinking with positive emotions, you can change your nervous system and improve coherence. Although for long-term change in most cases, you still need the work of self-reflection to understand why the pattern was there in the first place. This awareness allows you to familiarize yourself with what it feels like when the pattern is triggered, recognize it, bring in positive thoughts/emotions, and stop it in the future.

Electromagnetic Fields

Recent research shows that humans can sense magnetic fields and that people with certain genetic factors can experience increased sensitivity. We humans rarely notice field interactions as they remain largely unconscious. This is because they are subtle and often difficult to perceive through the ordinary five senses. However, these sensations are informed by our environment and the emotions of others. They are also a facet of our experience of empathy. For example, have you ever walked home at night and felt something tell you not to take a specific street? This is because the emotional fields of others project outwards as part of their biofield, and your nervous system can pick this up. Even unaware of these senses, the body unconsciously responds to these subtle magnetic fields.

From my observations, conscious and unconscious responses to magnetic fields are another layer of reactions to one's environment that affect the epigenome. Our brains and nervous systems take in and use a plethora of information in our surroundings to form responses, many occurring on the unconscious level. This, plus the library of our personal histories—such as how we react to previously experienced adverse life events—accounts for much of the lens we have created to perceive our realities consciously.

How Thoughts, Beliefs, and Emotions Affect the Epigenome

Senses are the nervous system's means of helping organisms interact with their environments and are closely tied to emotion. They are the brain's way of understanding sensory inputs concerning what is happening in the world around us. Such sensations can be interpreted as an approximation, not a truth, especially because past experiences can cloud emotional responses. This approximation profoundly impacts one's epigenetics as it creates its own "field" of information, which affects genetic transcriptions and, by extension, one's physiology. Furthermore, this means there is a direct connection between a person's consciousness and the DNA products (or proteins) that regulate the body's functions.

It is well established that prenatal stress and early childhood adverse experiences can expose epigenomes in the nervous system to emotional regulation issues. This can lead to issues such as trouble managing stress or even difficulties with addiction later in life. One study suggests that nightmares from traumatic events can even be epigenetically inherited. If a pregnant mother experiences high-stress levels, the baby's epigenome catalogs the outside world as stressful and shifts nervous system development to accommodate being born into a stressful environment. This has been observed to permanently and negatively affect the baby's immune system, metabolism, and brain development.

Strong, negative emotional responses take energy away from essential processes like healing physical wounds. For example, one 2008 study showed that when subjects were angry, stem cells did not strongly receive the message to cure their ailments because the body's resources were focused on building threat-response biochemicals. In such cases, the emotional signals essentially overpowered the repair signals; this demonstrates how information from the nervous system uses epigenetic changes to shift energy from one system to another via the regulation of DNA. It only takes one regulatory gene to drive inputs of potentially 1,000 other genes, and what informs that initial gene can be as simple as one thought, emotion, or belief.

Each person is a walking web of acquired beliefs from their lifetime. Many of these beliefs, however, are outdated and actively create a contradiction somewhere in our physiologies. As an adult, for example, do you still need the belief that you created at six years old about clowns being evil? Your mind quickly, within seconds, made that belief when you didn't feel safe, and it was then stored in your subconscious. The result is that the next time you saw a clown, which you had identified as a dangerous entity, you had a subconscious reaction that alerted your nervous system that you were not safe. This has practical applications when encountering dangerous situations, but we don't need to hold on to every past trigger. The problem is that these previously established beliefs never get consciously reviewed to determine if they are outdated or still needed, and the brain continues to retain obsolete beliefs. Two of the most common subconscious beliefs I encounter in my work are "I am not good enough" or "there is something wrong with me." These strongly internalized beliefs negatively influence one's perspectives like a fog clouding reality, making many life experiences more difficult and changing how one's genes transcribe.

A 2007 Harvard study by Dr. Alia Crum found that mindset and established beliefs may significantly affect a person's health. For example, one researcher invited a group of eighty-eight maids to a study. She told half of them that their work was great exercise; the other half was her control group and did not comment on them. Over the next thirty days, the first group experienced weight loss and lower blood pressure, while the second group experienced no changes. This exemplifies the power that mindset has on your physiology. While not specified in the study, I consider that this outcome was driven by epigenetic mechanisms fueled by a change in established beliefs.

Another study by Dr. Larry Dossey in 2000 was conducted around the effectiveness of prayer and belief in a loving god versus an authoritarian one. They found that the feelings associated with these practices can significantly affect an individual's healing potential. The positive emotions associated with care from others can also benefit the receiver and vice versa. The more you radiate a sense of friendliness, for example, the more you can also readily receive the positive energy of others via the interaction of biofields. Having such proactive beliefs also significantly increases one's experience of health and happiness in general. I often think of the old adage that asks, "What kind of universe do you live in—one that conspires to help you or works against you?"

The importance of mindset/emotions and the existence of human biofields is thus central to my work, as they are a key component of coherence, epigenetics, and how organisms interact. Each topic and concept introduced here could be its own book. The goal here, however, is to introduce you to the concepts behind my approach. The following section will present examples of applying this framework to my clients.

Bringing it All Together in Practice

When working with clients, I assess their current state and their goals. This could include genetic testing if desired, but all coaching is done from an epigenetic lens. I consider myself a caring guide who walks beside you on your path to wholeness. Many client reports feeling a deep sense of "being seen" and understood when working with me. My coaching has a holistic approach focusing on all aspects of one's lifestyle. The most successful clients turn the program's suggestions into a new way of being and understanding that becomes a daily practice. My current programs consist of weekly or bi-weekly sessions lasting three to six months.

First, we focus on lifestyle and mindset. These feed each other as an improved mindset creates and sustains the foundation for changes to your lifestyle. Your mindset must align with a belief in your abilities to learn and meet your goals. A subconscious sabotage mechanism will block your progress if this is not present. Since our minds can easily compile beliefs that are at odds with each other, I aim to help identify these "stuck" thoughts and shift them from a "fixed belief" into a "growth belief." Such a change can impact the emotional environment from which the brain shapes our reality; therefore, developing a "growth belief" can present new perspectives and bring empowerment.

Breathwork is the next fundamental component of my work. Unfortunately, our breathing patterns have become moderately to severely dysfunctional due to worldwide environmental changes, agriculture/eating habits, and conditions like chronic anxiety. You can, however, retrain the nervous system with breathwork. In addition, optimal breathing encourages a more homeostatic physical environment and a deeper connection with yourself. This is a realignment with a fundamental state of being generally lost in our modern society.

The final integral element of my work is the energetic level. I mix all the aforementioned practices with intuitive work, the Emotional Freedom Technique (EFT), homeopathic systems, and Theta Healing. Each modality works on the level of the belief system and emotions through physical tapping (EFT), reprogramming subconscious thoughts and emotions (Theta Healing), and frequency medicine (homeopathic systems). I also employ a specific kinesiology technique called Autonomic Response Testing (ART). ART allows me to gather information regarding your physical and emotional state by testing responses to your environment, foods, supplements, and more. If what is being tested causes stress for the autonomic nervous system, the client's ART response will show that by weakening the muscle being tested. A stress result indicates that what is being tested lowers coherence in the body. All sessions can also be done remotely. For in-person sessions, I have several devices to supplement programs. For example, I use a Healy microcurrent device and a Pulsed Electromagnetic Field (PEMF) therapy mat, which both use electromagnetic fields to improve communication in the body. The following case studies will refer to these as "Emag" sessions. I also utilize a Bio-Well device to track the coherence of organ systems throughout programs.

Case Studies

This first case exemplifies the general type of work I do in epigenetics and the success that can often be found, as most cases have similarities to this one but can take different directions.

Over a hundred clients have completed my programs and experienced life-changing successes similar to Taylor's. The second and third examples are from earlier in my career; they show some client successes but also highlight the need for consistency regarding goals and program length.

Taylor: A Classic Model of What is Possible

Taylor is a thirty-seven-year-old female with a history of Lyme disease; she has a caring nature and works as a nurse. The latent Lyme was activated by divorce and other stressful life circumstances dating back ten years. Her Lyme disease is currently well-managed through supplements; however, Taylor felt she had plateaued with her previous health practitioners and felt more was possible. Her complaints included fatigue two to four days per week, headaches, trouble sleeping, food intolerances (such as gluten and dairy), and bouts of constipation. She felt she had to do everything "perfectly" to keep her health well-managed. She feared that if she deviated from her diet or other practices that kept her stable, things would start to fall apart. Upon my first assessment, I could tell we had work to do on balancing inflammation and her lifestyle, mindset, and breathwork due to a lifetime of stress and perfectionism.

We first ran a genomic test to consider any epigenetic factors contributing to her existing complaints. It came back with suggestions on additional food modifications, such as cutting out glutamic acid sources, decreasing carbohydrates, and adding protein and fat. I suggested a paleo cookbook to help her eat more, in line with these suggestions in general. Her results also indicated difficulty converting beta carotene into the active form of Vitamin A, and I suggested switching to this form. I saw her again two weeks after she began making these changes, and she told me excitedly, "The bumps on my arms that I have had since I was a child are gone!" I told her that this was due to her body being able to utilize the new Vitamin A supplement, and this was helping with skin cell turnover and that it would also help her immune system.

Within two weeks after cutting out grains and other possible sources of glutamic acid, she reported having less constipation and noted that her brain fog had nearly disappeared. As she had improvements, we then used ART to identify additional foods that could be removed from her diet to improve digestion further. Chocolate came up as one of the recommended foods to cut; Taylor was not thrilled, but she admitted that she was using chocolate as a replacement for coffee and that there was also an emotional component to its consumption, as it was one of her comforts. I knew this wasn't something she had to let go of forever, but that refraining from it for a while would help balance her body and reduce anxiety. In addition to temporary removal, we worked on her beliefs about chocolate. She had a subconscious belief that "chocolate makes life sweeter." However, beliefs are usually layered, and it takes getting down to the seed of the "fixed belief" to make lasting changes. The seed here was that "there is not enough." I asked Taylor if, when she grew up, there was ever a lack of food. She explained that while there was never a complete lack of it, a scarcity mentality pervaded her family. We used Theta Healing to help replace the "fixed belief" with a "growth belief," She went home with mantras to practice daily.

Taylor returned in a few weeks and reported that after our last session and using the mantras for a week, her relationship with chocolate began to change. Now aware of where the perceived need had originated, the urgency to eat it diminished. ART showed that although her body's reaction to chocolate had improved, it was not yet time to reintroduce it. During this visit, I introduced breathwork techniques for her to begin practicing daily. This was an essential piece of the puzzle for Taylor because, like many people, her nervous system was in a heightened activation state. By better regulating her nervous system on this level, inflammatory response and sleep issues would potentially improve. We focused on breathwork for the subsequent three sessions to establish it in her daily practice. After these meetings, she reported experiencing a greater sense of calm, peace, and less anxiety—the latter she didn't even realize she had. Her sleep patterns had also begun to improve, and she was simultaneously experiencing increased energy.

We worked on Taylor's mindset and "energetic hygiene" in the following sessions. While we had already begun this process in previous sessions, getting the nervous system to calm down before deep diving into her mindset supported a long-lasting change. Previously I have observed that people who are both emotionally sensitive or nervous system activated with chronic infections, such as Lyme disease or the Epstein-Barr virus, tend to absorb the energies of others, mostly subconsciously. This trait can contribute to a personal feeling of disempowerment and a general energy drain that often accompanies the individual's medical condition and keeps them feeling unwell. Further, conditions with an autoimmune element often have a component of "not knowing self from other" on multiple unconscious levels. This may show up as overdoing or having super empathy for others. These people can have a quality of increased permeability to others' emotions and needs.

To keep Taylor's energy clear, she learned to create more healthy boundaries—not giving others too much energy and not taking on emotions from others. In her next session, she shared surprising insights about her dynamics with family members, co-workers, patients, and even friends. She couldn't believe how little energy she had reserved for herself in the past. She subconsciously believed that she needed to give much of her energy to help people, making almost every encounter a draining activity. Instead, she reported feeling more empowered with her newly established boundaries.

I had intuited a potential career change for Taylor some sessions prior, but it only became a focus in our last few meetings. This suggestion resonated with her deeply and evoked tears because she felt stuck and could not help people effectively in her current position; however, the transition felt daunting. We worked on breaking down beliefs contributing to her fear in this area, and she went home with mantras and some job transition ideas to consider. After six months, Taylor arrived at her last session as a changed person. Her original complaints had been resolved, and she noted she was experiencing more joy than she had ever thought possible. During this final program session, Taylor was still figuring out her career but, two months later, had begun study in an integrative medicine program. One year after, she left her job and found a position in a new integrative clinic.

By assessing and aligning to epigenetic principles on multiple levels, Taylor could start a life that she couldn't previously imagine; her symptoms had improved to the point that we could even reintroduce chocolate by the last program session. Taylor was one of my more motivated clients. She was open and willing to make profound changes on multiple levels and had high self-awareness. Although readiness and the speed of progress vary by client, everyone has their own timeline, which is perfect. My role is to be a guide who accompanies you, points out possibilities, and brings a different perspective. After finishing a program, clients often continue working with me monthly to stay on track with their goals or receive additional recommendations on adjusting their routines.

Nicole: A Case with Greater Implications

Nicole is a sixty-eight-year-old retired female who came in with a particular issue: she had severe pain after surgery on her left shoulder eight weeks prior. Over-the-counter pain medications weren't working, and she had experienced adverse side effects with prescription pain medications. She described her pain as an eight on a scale of ten and stated that her shoulders were achy, making it very difficult for her to accomplish normal daily activities. Throughout the first session, she clutched her arms and winced in pain; she said that was how every day was. I suggested a sixty-minute Emag session with me and a high-quality turmeric supplement.

She returned in a week and reported having felt moderate improvements in her pain since the first session; her demeanor was also notably much lighter. The supplement had arrived the previous day, and she had already begun taking it. Nicole had her second Emag session and later returned for another during the third week, after which she was doing markedly better and couldn't believe that her pain was nearly gone. A week later, she said that it had been completely resolved, and at her three-month check-in, she revealed that she was still pain-free.

Nicole had encountered an acute issue, which was an inflammation regulation problem after surgery. We were able to bring in support with turmeric and Emag therapy. Notably, Emag therapy has qualities that can support coherence when signaling is lost. There were likely a variety of mechanisms causing the signaling problem, including lifestyle and emotional aspects. However, she was not interested in a more thorough program to identify these causes at that time.

I include this example because some clients want to work on a specific issue, which is fine. Epigenetic principles still apply even if we do not tackle all underlying problems. However, this highly focused approach is only possible with acute issues; even then, more comprehensive work may be needed. For example, Nicole returned about a year after our last meeting with severe pain in her mid-back that had no known initiating factor. A sixty-minute Emag session resolved the excruciating pain, but I told her some additional lifestyle changes could improve her quality of life in the long term. As a result, she agreed to a three-month program.

Nicole had been a yo-yo dieter her entire life. For two years, she ate nothing but chicken to lose weight and also tried low-calorie packaged diet food. More recently—and for the third time—she had begun a stimulant prescription prescribed by her doctor and lost fifty pounds. However, as soon as she finished the medication, she regained it. Over the first six weeks, we applied all the epigenetic concepts from mindset, belief work, and lifestyle changes, and she lost seven pounds. We continued sessions for another six weeks to retain accountability and make additional adjustments to her program. After that, she continued and slowly lost thirty pounds over six months. I occasionally see her for what she calls a "pep talk," through which we realign her with attainable goals.

Jan: How Goals Drive a Client's Outcome

Jan is a fifty-five-year-old female with a gregarious and fun personality who works as an office manager. She was referred to me by her doctor. My role was to guide her on improving her eating habits to support healthier cholesterol levels and weight. Her routine at the time included biking and exercising at least four times a week. During our first meeting, she was upfront about not wanting to change her eating since she felt she already knew how to do so healthily. She also felt that her physician's recommendations were too limiting.

Additionally, Jan emphasized an important goal of hers: finding a healthy partnership. She was, at the time, in a complicated relationship with a man living in another state and felt it was neither healthy nor what she truly wanted. I told her we could work on this in conjunction with the food program. We first assessed where she was struggling with her cholesterol management, and I suggested a book with meal plans based on her physician's dietary recommendations. Though she was not thrilled, she was ready to put in the effort. At the same time, we worked on shifting beliefs that could have hindered her current goals.

In the next session, we checked how she was doing with her food goals. She noted that she had read some books but hadn't applied much information to her diet. We considered methods for encouraging more proactive dietary choices and employed ART to identify other foods that might have been causing a problem for her. Dairy and gluten came up in the results, so I advised her to remove them from her diet for four weeks and focus on moving toward a paleo diet. At the same time, she was excited to work on her relationship goals, so we employed the EFT and belief work on both topics simultaneously.

Jan returned one month later and reported difficulty staying with the assigned food changes for over a week. She knows she feels better when removing dairy and gluten from her diet but didn't want to. She resumed her relationship with her ex-boyfriend but felt that her understanding of their dynamic had shifted from our first session; this observation opened an opportunity for change. I made a few more suggestions on sticking to the dietary adjustments while working with her on her belief systems around food and relationships.

Another month passed, and Jan came in reporting that she had done well on the diet, was experiencing more energy, and lost five pounds in the process, but she confessed she also had consumed food I recommended she not eat. Additionally, she felt ready to let go of her current relationship "for good" and wanted to try online dating. We continued to work on strengthening her habits and developing new beliefs.

At her next appointment, she revealed that she was excited to be going on first dates with two people she had met online. She admitted that she was not as interested in my help with the food facet of our work as the other, as she had noted in her first session. After this was established, we focused on the personal relationship aspect of her life for which she wanted help. As a result, she was engaged in six months and married after three more!

As we saw in Jan's case, it is ultimately up to the client to determine their goals. I have so much fun working with all my clients and am ready to meet them wherever they are. I want everyone reading this to know that if you want to create or change something, you absolutely can! Our consciousness, however, must be prepared to make the necessary changes that manifest our heart's desire. In my sessions, I leave space for the client's process and timing because I have observed that sometimes the soul wants a particular experience in this life. We can free ourselves from obstacles, but it's completely okay if that choice isn't made. For every individual, this moment is a human experience that they get to create uniquely. I hold a loving space and support for it all, and I always feel honored to be part of peoples' journeys and celebrate whatever growth the individual is ready for.

The Hidden Powerhouse of the Mind

by Corinne Sullivan and Camille Sullivan

We are one system; the brain is integrated into the body at a molecular level, and therefore neither can be treated separately without the other being directly affected. Our bodies are, in fact, the realization of our subconscious minds.

Candace Pert, Ph.D.

Transforming Our Conditions

We all want to experience liberation and achieve a life of fulfillment free from fear and sabotage behaviors. But how does one overcome fear and personal saboteurs? This is where it can get tricky! One of the mind's primary tasks is to keep us safe. Change alerts the mind of danger. Therefore, our mind is constantly trying to halt any change from occurring. Moreover, when we run on former belief paradigms from our subconscious programming, we're not effortlessly able to address what we need to change and allow better conditions to materialize. The mind, simply put, doesn't like change.

Fortunately, there are powerful techniques to rewire the neural circuitry to effectively heal the distortions of our psychological programming and overcome sabotage behaviors. However, to successfully change our conditions, we must first unpack our belief programs and override our cyclical patterns to experience lasting, effective change. Otherwise, we will continue to recycle the same painful lessons and familiar patterns rather than gain freedom from our limits.

Cyclical Patterns

Whether experiencing resentment from unhealthy relationship cycles, going through a divorce, or having recurring patterns from childhood trauma, we seem to relive painful life lessons and vicious cycles. To varying degrees, these are likely the rigid patterns from *epigenetic inheritance*. Epigenetic inheritance refers to transmitting certain epigenetic marks to offspring and is often challenging to overcome. At least, not overnight. It means that a parent's experiences, in the form of epigenetic tags, can be passed down to future generations.

Perhaps you've done all you can to lose a few extra pounds, yet the stubborn weight doesn't want to shed. Maybe you've worked hard to get ahead financially or advance your career. Or you can't seem to accomplish your goals, yet you have done everything possible to get ahead. When negative cycles keep repeating in our life, it may help to peek under the hood of our psychological DNA to see if our relatives struggled with similar issues. Why? Because our inherent programming is generally the most difficult to change.

We tend to develop a preference for things merely because we are familiar with them. This is a psychological phenomenon known as the 'mere-exposure effect.' In social psychology, this effect is sometimes called the familiarity principle. This effect on the human mind can influence feelings and decision-making.

Why does the Brain like Familiarity?

From an evolutionary perspective, Raj Raghunathan, Ph.D., explains that familiarity can make us feel comfortable and safe because it reduces the uncertainty of anything that might pose a threat. Generally speaking, familiar things are likely safer than things that are not. Exposure to new or unfamiliar stimuli initially elicits a fear of uncertainty. If something is familiar, we have survived exposure to it, and our brain recognizes this and steers us toward it. The brain resists uncertainty as its most significant threat for survival purposes, although it's not necessarily safer. For instance, an abusive relationship can seem familiar, although it is unsafe and may have developed out of the "mere-exposure effect" from a familiar relationship.

Many of us are aware of certain belief systems that hold us back from success and happiness. However, many may not be aware that most of our core limiting beliefs and recurring patterns are genetically transferred into our psychological programming, a process known as "Transgenerational Epigenetic inheritance" (TEI).

Much of the distress we experience in life results from a genetic or an epigenetic predisposition because it was genetically inherited from our parents, grandparents, and/or other ancestors. The research of Jablonka and Raz found that TEI increases our susceptibility to not only diseases and pathologies, such as physical health conditions, but also the psycho-emotional transmission of our ancestors. This means that even the transmission of traumatic experiences from previous generations is passed onto later generations because a traumatic experience produces epigenetic changes that can be transmitted.

Ask yourself what significant hardships and challenges your ancestors faced in their time because most of our stress-related symptoms and other psychological challenges were transferred in our DNA from one family line down to the next through RNA strands. These genetically transferred stresses and hardships can lead to built-in propensities toward psychiatric conditions and poor coping mechanisms, such as stress, anxiety, panic attacks, insecurities, self-doubt, anger, depression, self-sabotage, addictions, and other unwanted physical and behavioral issues. That pretty much sums up most of our underlying challenges!

Psychology of Beliefs: Our Mind's Powerhouse

Trans-Generational Epigenetic inheritance (TEI) has also provided a valid explanation of how our beliefs and subconscious programming are influenced by the primitive beliefs of our ancestors' memories and life experiences that shaped their beliefs and transference onto future generations. This means we're also genetically wired or subject to some of the same inherent limiting beliefs and negative subconscious programming of our ancestors due to some of the hardships and traumas they may have endured.

Therefore, subconscious beliefs have a major influence in preventing us from creating what we truly desire and, instead, often resulting in self-defeating behaviors and sabotage cycles—all of which may have originated from these inherited and imprinted beliefs. This is known as Inherited Subconscious Programming.

Epigenetics reads the instructions to switch our genes on and off. What determines the switch is our subconscious programming. Subconscious beliefs also run our psychological programming because our memories, emotions, and beliefs influence our thoughts, choices, and behaviors. Even if you don't know your biological parents, subconscious beliefs are encoded in your genes and primarily determine the actions that make up your reality. The subconscious mind is the star of the show!

As we form beliefs, our beliefs form our reality—consciously or unconsciously. Therefore, our emotions primarily create our reality and form our beliefs.

Once we shift internally, we can then shift externally. Therefore, if we want to improve our circumstances, we must first strengthen the *genetic* psychology of our beliefs because our personal psychology reigns supreme over our circumstances.

Each of us realizes that we are, in part, a product of our parents. Yet we may not be aware that our inherited genes dictate our outcomes, including our successes and failures. For instance, we repeat our parents' lessons or familial patterns, whether marital, financial, or other challenges.

We do not want to repeat our parent's mistakes; therefore, many of us over-correct. Yet, when we attempt to run away from our past and resist becoming like our parents, we meet more resistance and invite opposition into our lives. However, when we correctly recode the programming, we can establish a healthy balance within ourselves and achieve our desired goals without sabotaging our efforts.

Epigenetic inheritance can be responsible for the same programming and similar realities that influence our belief system, even common subconscious programming and limiting beliefs that distort our perceptions and lead to sabotage patterns. Common subconscious programming and limiting beliefs such as:

- Believing you have to work hard to succeed and struggle to survive.
- Not believing you're good enough and worthy to receive more.
- Insecurities and ego-based fears, including jealousy, competition, unhealthy entanglements, or other power struggles in relationships.
- Conflicting beliefs, for instance, "I want to be fit, but I don't want to eat healthily and exercise."

These belief systems result in repetitive negative cycles and self-destructive behaviors. Yet, there are solutions to recode epigenetic patterns for everyone and every life circumstance.

Self-defeating behaviors are familiar to everyone. Fortunately, we can use effective strategies to reprogram our subconscious mind and recode our limiting beliefs to overcome these stubborn genetic propensities, enabling us to thrive. However, because science has proven that genes are not only the physical characteristics of our lineage but also unresolved emotions from traumas transferred from our ascendants, our epigenetics may be the most significant cause of all our saboteurs.

Fortunately, we can signal the correct gene code to manifest desirable outcomes because our genetic expression is changeable. With that said, did you know that healing our epigenetic inheritance is the fundamental step to unlocking our unique mind power and DNA potential?

Our Story

We were repeating painful lessons that led to additional hardships throughout our life. No matter how much we consciously and subconsciously worked at it, the transgenerational cycles continued to operate. That is until we addressed the root cause of our psychological programming by utilizing a specific epigenetic recoding technique, which requires nothing more than accessing the hidden powerhouse of the mind—on a genetic level.

As twins, we experienced trauma at birth and subsequent adverse childhood experiences, including childhood trauma and autoimmune disorders that progressed to other chronic health conditions. Because we are all drawn to what we need, this led us on the healing path to remedy these conditions using only natural approaches and to address the other adverse experiences in our youth, which we will get into later.

Medical professionals didn't understand the origins or pathology of these various conditions — let alone educate us on how to correct them. Years of being on the medical merry-go-round without success only took us further off course, causing us to feel defeated. The jury was still out; however, we knew we didn't want to treat the symptoms alone when it was a holistic issue because we wanted the whole equation, which compelled us to find the actual cause and panacea.

Health Challenges

Corinne was diagnosed with Fibrocystic Breast Disease (FBD) at age eleven, otherwise known as chronic cystic mastitis. Unfortunately, doctors failed to inform her that this condition was associated with a reproductive and potential autoimmune disorder called Polycystic Ovary Syndrome (PCOS) and often co-occurs with endometriosis.

PCOS (Polycystic Ovarian Syndrome) is an inflammatory condition and is the most common hormonal disorder among reproductive-aged women, usually caused by an over-production of estrogen or xenoestrogen toxicity. Women with PCOS have ovaries that create an abundance of follicles (FSH) each month without producing an egg. As a result, no single follicle becomes dominant, so ovulation will not occur as usual, and estrogen and androgen (testosterone and DHEA) levels remain high, causing symptoms. PCOS can contribute to irregular periods, depression, excessive weight gain (despite diet and exercise efforts), excess male hormones, acne, excess facial hair, polycystic ovaries and breast tissue, excess insulin, and infertility.

Endometriosis is a disease characterized by tissue resembling the endometrium (the lining of the uterus) outside the uterus, causing pain and infertility. In addition, it causes a chronic inflammatory reaction that may result in scar tissue (adhesions, fibrosis) forming within the pelvis and other parts of the body. Endometriosis is often a chronic disorder associated with severe, life-impacting pain during periods, sexual intercourse, bowel movements and urination, chronic pelvic pain, abdominal bloating, nausea, fatigue, and sometimes depression, anxiety, and infertility. It is a disorder that most commonly involves the ovaries, Fallopian tubes, and the tissue lining the pelvis. When endometriosis involves the ovaries, cysts called (known as) endometriomas may form.

The Department of Obstetrics and Gynecology, Faith University School of Medicine, Ankara, Turkey, conducted a study investigating the association between PCOS and FBD. The conclusion of these studies showed a statistically significant association between the two conditions. Studies further report that women with fibrocystic breast disease should be evaluated for polycystic ovary syndrome and vice versa. As identical twins, we were prone to very similar genetic dispositions, and evidently, we both endured severe symptoms of PCOS and endometriosis. A double whammy!

Of course, no doctor, gynecologist, or nurse further evaluated this condition. Instead, we suffered for many years, pleading year after year for help to correct the condition or, at the very least, alleviate some of the symptoms we both suffered. The nurses and medical doctors couldn't help; this was a novel condition they hardly understood, nor could they offer any direction or advice on treating it without recommending drugs or surgically removing vital organs. Surgical removal of the uterus (hysterectomy) and ovaries (oophorectomy) was considered the most effective "treatment" for endometriosis at the time. Unfortunately, this surgery can lead to early menopause if you are a young adult.

We were continually referred from one specialist to another instead of being seen by a single physician who could understand or enlighten us on the interconnection of our anatomy. This seemed very limiting and never provided us with any correct answers, leading us further away from the solution. We were exhausting our resources and lost faith in the medical system. It was evident that the traditional medical route wasn't our answer. We didn't feel it was wise or intuitive to resort to harmful drugs or risky surgeries in an attempt to solve our health problems on a holistic level. Instead, we wanted a practical approach to addressing the source and etiology of the issue. What initially caused PCOS and endometriosis, and how could we successfully treat it?

It wasn't until meeting with a Naturopathic Doctor who could make a complete diagnosis and had a broad knowledge and method to correct some of these conditions that our lives changed—for good. Apparently,

Naturopathic Physicians had been successfully treating endometriosis and PCOS all along! Consequently, this led us to pursue an education and career path in natural medicine as holistic practitioners and therapists, and we remained in practice for nearly two decades. Learning about natural sciences inspired us to adopt a holistic lifestyle and maintain a healthy standard in our personal and professional lives. Our philosophy was not only to practice our own medicine but also to *master* it.

As we cultivated a series of different healing modalities, we discovered the solution and pre-eminence in holistic sciences. What we found most appealing about holistic medicine is that it is centered on addressing the *whole* scope of the individual on all levels; mental, physical, emotional, and spiritual, achieving total wholeness - leaving nothing out and pinpointing the root cause. Knowledge of these natural sciences could have saved us many years of pain and suffering. However, it would have prevented us from finding the answers we longed for and coming full circle to the solutions we were about to discover.

Due to PCOS and endometriosis, neither of us could bear children– at least not safely, and at best, we would be high-risk. Therefore, we would primarily entertain romantic relationships either being fully invested or primarily with divorced men who already had their own children. We did not want to deprive any men of their desire to have children. However, settling down was challenging, particularly without seeing the sabotage behaviors. It became evident that we were "settling" in our romantic relationships rather than settling down with a long-term companion or spouse.

Marital Patterns

At the age of 33, Camille settled down in a relationship. However, many challenges indicated potential long-held issues, specifically the polarized pairing of a narcissistic man who started leading a dual life and escaping to addiction instead of addressing his trauma with an empathetic woman who found herself following in her mother's marital footsteps quite rapidly. Familiarity? Although Camille did her best to correct her generational patterns, the natural propensity to magnetize familiar experiences felt like a tug-of-war in her mind.

Within five short years, it was conclusive that Camille had married into "familiarity" with a man that didn't initially share any semblance with my father. However, she knew it was due to her "parent-interaction" patterns. We unwittingly transfer our patterns onto our spouses or significant others. Because of the unhealed aspect of her youth with her father, Camille had attracted the paralleled version of an unhealed man, and the only thing holding them together was a "trauma bond" similar to her folks.

They had no hope because neither was unwilling to continue the "high-low" rollercoaster ride. Nothing in the relationship was meritorious to a potentially healthy, long-term marriage worth fighting for. As Camille walked away, she reflected on what she had learned. The situation would only worsen over time. She witnessed the man she had erred in believing deserved her all take daily nose dives. He was headed down dangerous and unforgiving territory and appeared to have a death wish. On the other hand, Camille was ready to engage in better ideals and shift gears into a higher trajectory. Camille wanted much more for my life.

This story illiterates that our incongruent values and lifestyles placed us on very different paths. Both of us subconsciously knew that our inability to reproduce was also our body's natural intelligence, which is the unconscious mind, communicating to us *not* to continue passing our generational patterns onto future offspring, or at least until it was healed. Our body is one of our best teachers because our physiology manifests our unconscious mind.

Childhood Adversity

As twin sisters, we grew up in a hostile environment with a dominant and narcissistic father that would generally be described as physically and psychologically abusive. Our father had genetic tendencies to behave with anger, violence, criticism, domination, control, and substance abuse. Although our father was very difficult, with an authoritarian parenting style, later in life, we learned that he endured the immense weight of trauma and family loss that he carried with him throughout his life but regretfully never addressed. It was clear that dissociating from his problems was his poor coping mechanism to survive the emotional pain of his past.

Consequently, his unresolved trauma left him enraged. It led to an unhealthy projection of his anger misdirected at his wife and children while escaping addiction and other destructive behaviors. The one who was at the greatest harm to this abuse was our sweet mother. She became a victim of his violent temper, and there was domestic violence toward her and her five daughters because of her passive personality. She had a strong and wonderful mother, yet her father also had a harsh temper and was guilty of frequent escapism to alcohol. Our mother also married into familiarity. After all, this was normal to her because she didn't know any different.

As Camille and I matured, we possessed great compassion for all the hardships our father had endured and took accountability to heal our relationship with him. Naturally, we had no desire to wind up with anyone who reflected his unresolved trauma that resembled a masked villain. We knew we deserved more than that and didn't want to be relegated only to gene expression! So, we swung the pendulum in the other direction.

Sure, our father had good qualities too. However, we both had made a concerted effort to achieve higher self-awareness and reprogram our

DNA to ensure we were consciously aligning ourselves only to men who single-handedly possessed his good qualities.

We connected so many dots to these unconscious patterns, chalking it up to–our epigenetic expression! Resentment festers. With divorced parents and other siblings who had also undergone divorces themselves for similar reasons, falling in the same footsteps by marrying men like their father (their unhealed relationship with their father), the marital pattern was inarguably encoded in our DNA. We learned first-hand that our psychology and epigenetics are responsible for most of our outcomes.

Naturally, we wanted to unpack our subconscious mind and overcome any remaining beliefs and subconscious programming leading to less than copacetic outcomes. Most people don't go deep enough to change their limiting beliefs subconsciously, let alone on a genetic one. It felt like a challenging feat. However, we weren't ones to give up easily, and our strong suit was to transform our conditions completely. We knew we were onto something and needed to explore the research deeper.

Transgenerational Transmission of Trauma (TTT)

To lend a unique illustration of epigenetic patterns and a personal example:

Corinne suffered an injury in early adolescence, causing her to lose the same teeth as our mother's parallel childhood injury, which resulted in her losing the same teeth in her youth as our mother did. Ironic? Clearly, separate teeth injuries are not genetic but a coincidental transgenerational pattern. We carry the same psychological burdens from cognitive distortions as the generations before us. They appear in our lives as similar patterns that include emotional and physical dispositions such as epigenetic inheritance or TEI.

Another example and case study of epigenetic patterns from transmitted trauma relate to our father, who suffered a heart attack just as he was approaching his 50th year. Likewise, his father suffered a heart attack in his 50th year, and his grandfather died of heart disease.

In addition, after 20 or more years of working with clients, we discovered the common thread and found that epigenetic inheritance was at the heart of everyone's challenges. We went on to integrate epigenetics into our practice and collaborated to address various combined, holistic approaches with our clients. We addressed trauma, addiction, grief counseling, terminal and advanced stages of cancer, and virtually every known mental, emotional, physical, and degenerative ailment.

Each of the natural healing modalities and disciplines we implemented was very instrumental in making positive strides. However, we effectively overcame these core challenges when we took it a step further to heal our psychology at the core genetic level. We addressed the root of our epigenetics inheritance.

Not only have we witnessed successful outcomes in the lives of our clients with these therapeutic methods, but we, too, have proven the efficacy of these holistic healing systems and the ability to overcome the negative consequences of TEI.

We dedicated our entire life's work to this professional vein and became our family's chain breakers, healing our epigenetic inheritance. We discovered the winning formula and final breakthrough method, which led us to develop our unique 7-step P.A.N.A.C.E.A. Model and design our advanced healing system. The 7-Step P.A.N.A.C.E.A. Model is an internationally accredited personal development course devised to overcome sabotaging behaviors, childhood trauma, and generational patterns by re-coding the subconscious mind on a genetic level, getting to the source of the issue and belief system! It contains a therapeutic workbook, self-analysis techniques, step-by-step modules, and strategic Genetic Recoding® exercises that lead clients to their own answers to recode negative and genetically inherited programming and achieve greater liberation by removing limiting beliefs and generational cycles. Genetic Recoding® is a holistic and comprehensive therapeutic process that naturally re-codes negatively inherited and indoctrinated programming, fears, and sabotage behaviors. It activates epigenetic potential, starting at the root of core issues and genetically transferred dispositions. It addresses core subconscious programming on a genetic level, including childhood conditioning, inherent self-defeating, and relationship patterns. It also incorporates generational healing as a chain-breaker from genetically transferred dispositions and other transgenerational-related issues. These programs were created for both personal and professional practice as well as for individuals, couples, families, and professionals to liberate future generations moving forward! It's an extremely effective and innovative approach to family therapy. After designing therapeutic Genetic Recoding® programs for our clients and witnessing their powerful breakthroughs, we decided to make them available on a larger scale through online practitioner certification.

Mandy's Epigenetic Relationship Patterns

Corinne's client, Mandy, struggled with co-dependence (attachment disorder) because of her feelings of abandonment as a child. Her mother never married her biological father, and when Mandy was about age five, her mother married a wonderful man who was a great father to her—someone she felt really loved by and close to. However, her stepfather, whom she dearly loved, passed away when she was a young teenager. As a result, her mother became unstable in relationships and always brought home a different man. Mandy resented her mother's immoral behavior and never found someone to replace the closeness she had with her first stepfather. She felt that her mother lived an irresponsible life and vowed she would never be like her mom or live such an inappropriate lifestyle that destroyed her kids' lives and stripped them of their childhood innocence. Oh, that *resistance-persistence* principle!

In her early 20s, she married and had three children. She wanted so much for her husband to love her the same way she loved him in return and deeply sought his affection. She did what it took for him to desire her or "get his love," even giving herself away sexually when she was yearning for emotional closeness. He couldn't give her that, so she settled for only what he wanted. Over the years, she felt that she had become a "sex object" to him and that he only expressed his love by reaching out to her physically and sexually. This led to dissatisfaction and a lack of fulfillment in their marriage. She began taking control of the situation by withholding and refusing physical intimacy with her husband whenever she felt necessary. This tactic (just as a child who manipulates) was ineffective, and her husband began to cultivate an addiction to pornography, something he had always had a curious temptation with, which was his way of controlling and withholding in return.

Instead of finding the love and mutual intimacy they sought and needed, they rejected one another, deteriorating the relationship. The marriage finally came to an end resulting in a divorce. She was reintroduced to the life of being single and dating again. So far, her marriage was the only relationship she had ever had. Once she began dating, a pattern emerged as she repeated the same dating experience repeatedly. Now, she had more relationships to compare her patterns to.

She immediately flirted with almost any guy she could get attention from at work. Once again, back to the beginning pattern that she learned from her mother (who suffered from the grief of losing a spouse), she had learned that sexual attention is better than loneliness or no attention at all. She used this method and found herself with multiple partners. Unfortunately, shortly after each partner, the "relationship" seemed to end only after a few weeks, and the guys only called when they wanted her physically. She eagerly clung, pursuing after each of them, and was so angry and hurt that this was all they wanted. She couldn't understand why none of them wanted a committed or serious relationship with her.

She decided she couldn't trust men because they only wanted one thing from her. She wanted to be loved for who she was rather than objectified, so she continued multiple physical relations. She was constantly disappointed that she still could not find one that offered her the "real thing" and wanted a commitment to her. She persistently urged, clinging to each of them, and tried desperately to interest them in marrying her. This only pushed them away, and the letdown continued repeatedly.

She still did not recognize the origination of her behavior patterns. She did not realize that she was repeating her mother's behavior for fear of becoming like her (projecting her fears). She also did not see the pattern in her co-dependent behaviors from her fear of abandonment after losing a real father figure she felt had left her. Additionally, no real or stable father figure ever returned to her life. On a subconscious level, she had convinced herself that she was unlovable and that no man would stay with her. Only by entertaining cheap physical romance with men would she get their attention, and she would rather have sexual attention than no attention at all. This way, she at least felt desired. But in the end, she still felt defeated, empty, and alone.

Corinne began counseling her while subtly pointing out her patterns. She understood that continuously having coitus relations with men to "get them to like her" was unhealthy. She agreed to have chaste relationships and seek more quality men than she had been. However, Corinne pointed out that she was equally using them in return, so how could she be angry with them for doing the same to her?

Corinne suggested that Mandy avoid dating for a short while and develop a more loving and forgiving relationship with herself while recognizing her codependent and abandonment behaviors. She gave up dating for over a month, which was only frivolously short-lived. During that time, she began to see herself in a new way. She started to feel a greater love for herself and felt strong enough to say that she didn't need anyone to help her feel loved. She began making headway; however, patterns take time to overcome, and it's a step-by-step process—two steps forward and one step back. She then met someone she was interested in.

This one was different and did not try to have a physically intimate relationship with her. After three weeks, she was convinced she was falling in love with him and could envision them potentially getting married. Her behavior improved, and the two found a mutual liking for one another. However, he had some hesitation throughout this time and did not define an exclusive relationship with her. She had moments when she became paranoid that he was cheating on her (even though an exclusive relationship had never been established) and that maybe he wasn't interested in her after all. She would quickly withdraw from him in fear of him leaving her. This is a common pattern of letting fear and pride get in the way of relationships. Corinne continued to point out that she was acting out and projecting these fears of abandonment onto the relationship and that these issues needed to be addressed. Anytime she let go, felt her true inner feelings, and loved herself through it, she found that he would immediately prove himself to her and once again convince her of his feelings and affection toward her.

Things between them would carry on this way, and she would slowly begin to improve. When things started to go well, she began to obsess over the relationship, and a greater fear of him leaving followed. She so desperately wanted to feel loved and accepted by him that she placed him unequally above her as a superior. Out of desperation, she would push harder and cling tighter, only resulting in pushing him away because of her fear due to her childhood abandonment. Seeing that he would distance himself due to her pressure on him, she would make impulsive decisions and assumptions toward him, finding reasons to sabotage the relationship. Once again, fearing him leaving her, she would rather leave him than be hurt. She allowed her pride to take over to prevent herself from being convinced that her fears were genuine after all..."she was unlovable."

The relationship had much potential, but Corinne was unsure it could go the distance, especially when insecurity and pride took over. Corinne would counsel her not to have any expectations of the relationship. This was a great lesson or stepping stone for Mandy in recognizing and overcoming her patterns. Not every relationship can "be the ultimate one," but it does have the potential to discover your true self. The best and most obvious way to overcome our patterns is to see them in a relationship setting. This can be difficult to explain to a person with codependent patterns.

It seemed Corrine's advice did not satisfy Mandy when she only wanted to believe that he was the right one for her. As a result, he never felt comfortable or ready for a committed relationship with Mandy. Once again, Mandy was projecting her issues (fears and beliefs) into the relationship, unable to recognize that she was the one doing it.

So long as we continue to assume that others will hurt us in one way or another, they can never prove otherwise because we'll interpret everything they do negatively through our own lenses and limited perspective.

Had the couple stopped projecting their patterns onto one another, the relationship could have gone the distance. The power of our programming can cause us to overvalue something we fear losing, i.e., being in a relationship versus valuing ourselves and who we're in a relationship with. By potential gain or the fear of loss, we will see such programming revealed where we might hold too tight of a grip instead of freely allowing nature to take its course. Everything happens according to our beliefs and what we put faith in. This is often indicated by experiences in our youth (before age seven), which become imprinted in our subconscious. A child's subconscious mind is an open, absorbent sponge that embeds many experiences.

Regretfully, Mandy's experience is not unique. Generational patterns are common to all of us. Still, we don't always have the awareness to see them or the solution to overcome them and find ourselves repeating painful lessons instead. No matter what challenge, we are meant to use the resistance we face as a stepping stone to catalyze our growth potential.

When we *start* with our epigenetics, we quickly get to the root of the matter, and the results can assist us in the long term. When we first address our psychology, we gain insight into the "whole" scope of the individual (mental, physical, emotional, and spiritual) by starting with *soul*-level healing, which is the internal workings of the psyche since the body is not separate from the mind.

The root word of generational is gene, meaning we must *begin* at the *root* with generational healing. If we are to overcome obstacles and achieve lifelong success to advance our potential, instead of delaying it with needless suffering and living with regrets, it is vital to start at the "root."

Liberating the Family System

One of the most incredible benefits of doing our core generational healing is liberating our children from the ongoing burdens of these intergenerational chains. In this respect, the entire family benefits. With our collective years of experience, we co-authored and combined our healing solutions to move into this more profound work.

When we operate from the fear-based programming of the ego, we will stay stuck in familiar "comfort zones," causing us to resist change. It's been said, "Comfort is the enemy of progress." With our 7-Step P.A.N.A.C.E.A. system, using empowering psychotherapeutic strategies, we can dissolve the ego's self-defeating behaviors by genetically recoding the subconscious mind on a genetic level and overcoming sabotaging behaviors such as childhood trauma. This materializes positive change more rapidly for our client's growth potential. But unfortunately, if we prolong our deeper inner work, we will only continue to pass these predispositions on to future generations instead of creating fuller life of true, liberated living.

Our Mission

Our mission is to guide others in avoiding perpetual lessons and hardships. We encourage others to invest in a brighter future by applying lessons from our own experiences. Our program is designed for those who would like to overcome epigenetic programming and advance to the next level of their life. It is also intended to certify and empower other healing professionals. We're delighted to bring solutions forward, share with those ready to gain the same benefits and achieve results as we and many others have, and expand our reach to impact more lives.

Our goal is to empower anyone toward their self-mastery and gain victory over their genetic predispositions while rewiring their mind for personal success and finally *begin to live life by design, not by default! True Liberated Living.*

The Epigenetics of Longevity

by Melissa Petersen, DC,MS

Be Ageless – Live Limitless.

Melissa Petersen, DC,MS

What The Future Holds

Did you know at this very moment, according to scientists, the first person who will live to 150 has been born? In addition, eyesight has been restored in mice using cellular reprogramming, organs are regenerated in the lymph tissue, and gene editing eradicates disease. These are just a few ongoing longevity breakthroughs and advancements.

In the immediate present, we now know how to slow down and reverse biological aging. Different from the number of years on your birthday cake, biological age correlates with the rate and pace of aging in your body's organs, systems, and tissues. This is the main factor impacting how fast or slow your body breakdowns, dysregulates, or becomes impaired in its functions. Biological aging is a measurable marker that we can improve, and as it is improved, it reduces the rate of all-cause mortality, sickness, and disease so you can look, feel, and live better longer.

This is just the beginning. Thanks to the longevity field, the future holds a totally new reality. One is free from aging and decline as we have once known it. Instead, new potentials and possibilities exist where 100 years of age will become the new 50.

All of this is happening and being influenced by and because of epigenetics. Epi-above / genetics- the genome. This science looks at how the outer world and all the inputs and signals of life inform, influence, and impact the very expression of our DNA. It's within the DNA that our longevity potential is influenced epigenetically to either speed up or slow down the aging process.

Are Genes Your Destiny?

Growing up in a European and Italian family, I have the fondest memories of Sunday dinners. My Italian relatives would be around the table, playing cards, eating, drinking, laughing, and taking each other's money. As a young child, I vividly remember looking at my parents, aunts, uncles, and grandparents, who were full of life and vitality across the decades, and thought I would age well; I had good genes.

I held that belief until my mother passed away days after her 54th birthday.

If our health is tied to our genes, will my mother's fate be mine? Is there more to the story? This was a defining life moment that, little did I know then, would put me on my path to learning the science of epigenetics and redefining the understanding of aging as we know it.

Over time, research revealed the answer to my questions. First, when the human genome was mapped, it was believed yes genes are our destiny. Each one thought to cause a unique outcome.

Yet theory quickly unraveled when the same code sequence did NOT always produce the same result from person to person. Only a small percentage of the time was this the case, so what was at play?

They realized there was more to understand. Epigenetically, it wasn't the code (the underlying gene). It was how it was signaled or expressed. This expression from above the gene informs and influences it, essentially instructing it on how to respond. That was when it was realized that genes are our potential; they are NOT our destiny. Based on the signal, this will determine if and how the gene expresses itself.

Understanding Your Code of Potential

The genetic code is our individual blueprint, a recipe. It's the code of life that makes you, you, and me, me. And for all our individuality, you and I are 99% identical at the genetic level. The one percent difference that allows for all of our bio-individuality is due to epigenetics. The phenotypic expression of the gene is due to how the gene itself is signaled, informed, and instructed.

So, wait- does that mean there is a 99% chance that genes impact our destiny and that my mother's fate could still be mine?

No.

What research has validated time and again is that 75-90% of what determines how your DNA, your blueprint, will express into health or disease is lifestyle driven. So take a moment, and please let that fully sink in.

In the western world, we are marketed to every day to buy the next lotion, potion, treatment, or pill that will make us healthier, happier, more successful, more attractive, and more fill-in-the-blank. The truth is that 75-90% of what it takes to live our healthiest, happiest, and most fulfilling lives do not come in the form of a pill or potion. It's in YOU, influenced by your perception, beliefs, environment, and lifestyle choices.

Lifestyle drives this whole experience. How you live, where you live, whom you live with, what you are exposed to, and what you perceive and believe to be true, paired with your response or actions, are all variables talking to your genes. Each choice, each input, and each exposure is 24/7, informing and influencing your DNA to either express or repress your innate potential.

Inputs and Outputs

The quality of the input influences the quality of the output. It is this interplay that determines how our genes are expressed.

You and I live in the exposome; think of life as this bubble, and everything within it is talking to your body, cells, and DNA. Inside this exposome bubble or container, the signals from our daily lives, environments, habits, choices, thoughts, perceptions, beliefs, foods we eat, the air we breathe, relationships we are in, etc., all act as ingredients that interact with your DNA to either express or repress your innate qualities, capacities, and traits.

The signal comes in, saying, hey, life is good, it's safe for thriving, we need to replicate, reproduce, repair, create. This is the process of acetylation. Acetylation is a critical epigenetic modification that changes chromatin architecture and regulates gene expression by opening or closing the chromatin structure. Or it says it's bad out there; we must lock down, close off and protect, which is methylation. Methylation involves attaching small chemical groups called methyl groups, each consisting of one carbon atom and three hydrogen atoms, to DNA building blocks. Methylation isn't bad; it's an essential function. Your DNA must be closed and protected, so the code is not compromised. Yet, when we need to access the code, it is critical that it can open and be ready for replication. The code cannot be accessed easily and as needed when the body over methylates.

Methylation will increase in times of too many demands on the system. More and more protection is underway with less and less of the code available. Meaning the body is unable to access its greatest strengths and potential fully. It's literally in conservation mode, doing its best with limited resources. It is working harder vs. smarter, against vs. with itself. Methylation marks of aging across the body's tissue estimate your biological age and the rate at which your system is aging. By studying changes and patterns in DNA methylation over time in various body tissues, we can assess the rate of biological aging.

What this means in everyday life is that you look tired, aged, and drained. You feel exhausted and depleted like you can't keep up. You are not healing quickly, you get sick more often, and you feel off, out of step, out of sync, agitated, foggy, and even depressed. It's as if you are spinning your wheels and getting nowhere fast.

This indicates that our psychology is talking to our physiology. The outer world is talking to the inner world. Seventy-five to ninety percent of what happens internally is driven by the environment and interactions you face daily externally.

Is Your Body Working Harder or Smarter?

Everything is coming in, and the system (body) must interpret the demand, what it means, and what it needs. Is it helpful or harmful? Our body's job is to keep us safe and alive.

Your system is constantly assessing and determining if you need to close down methylate, protect and close off, or is it safe to activate, replicate and thrive forward? What's interesting is that there has to be a way, from a hereditary standpoint, to keep the species safe and alive and propagate generation after generation. Changes in the phenotypic expression of the DNA, based on methylation marks, are how the body passes messages down quickly from parent to child.

What demands do you place on your system physically, mentally, emotionally, chemically, electronically, environmentally, financially, and spiritually?

Checking in is the fastest way to tell if your system is overloaded.

How do you feel? Do you feel energized, clear, happy, and engaged in life? Are you vibrant? Do you heal quickly, sleep intensely, play full-on, and enjoy your life? Or are you drained, depleted, anxious, overwhelmed, overloaded, sick, or dealing with aches, pains, worry and doubt?

Demands are Stress to the System

Stress is neutral, yet, depending on your perception of it (the meaning we give it), the frequency, intensity, and duration of your exposure to it paired with your response all add up to tell your body if this is a "good" (Eustress) or "bad" (Distress) to the system.

Distress adds to an allosterically overloaded system that is locking down the DNA, over-methylating, working harder versus smarter, and zapping your quality and length of life. Distress increases your rate and pace of biological aging.

The autonomic nervous system is divided into two parts: SNS and PNS. I refer to the sympathetic nervous system (SNS) as the distressed state. It is when our foot is pushed down on the gas pedal of life—always going, accumulating, and draining the tank of the last drop of gas. On the other hand, when the body is stressed, the SNS contributes to what is known as the "fight or flight" response.

Distress activates your sympathetic nervous system, increasing cortisol, the hormone that increases glucose and insulin. This, in turn, impairs leptin, causes fat storage, insulin resistance, low thyroid, reduced sex hormones, reduced melatonin levels, and is pro-inflammatory. This is inflammaging at its best. Inflammaging is chronic low-grade inflammation occurring in the absence of overt infection. If this is ongoing, it biologically ages you epigenetically, affects your immune system, and other body system breaks down faster over time.

Eustress supports the parasympathetic nervous system (PNS), the second part. This is the rest, digest, and connect system. This is where the body can increase its own capacity to grow stronger and more resilient. A key component of Eustress is rest. Balancing the system while integrating this learning is executed best by activating the vagus nerve. This long nerve runs from your brainstem down to your abdomen. It is responsible for many different functions, including heart rate, digestion, and immunity. It is also one of the most important nerves in the body for maintaining mental health.

This nerve is the superhighway of the autonomic nervous system. When activated, it is like pumping the brakes on the sympathetic fight or flight distress pathways to engage the rest and reset pathways of eustress. You can strengthen your vagus nerve by breathing deeply and slowly, connecting with nature, and eating a whole-food diet.

When the body is energized, the DNA activates, and powerful hormones slow the rate of aging, thus optimizing healing, health, and ability to thrive. Simply slowing your breath rate, humming, singing, laughing, or even gargling can activate this powerful pathway and help you pump the brakes, moving from distress to eustress. You can also strengthen your vagus nerve by connecting with nature and eating a whole-food diet.

By letting your foot off the gas long enough to create micro-moments of rest and recovery, the system can "catch up." Sometimes, it only takes a few minutes to recalibrate so you can go further, faster.

Activating the Fountain of Youth

While no lotion, potion, or pill at this moment will magically erase a system in distress, there is a path to reverse age that can be like drinking from the fountain of youth.

As I have said, 75-90% of what will slow down and help us to reverse age is a lifestyle, yet, let's get specific. What would give you the most significant return on your health and longevity investment if you had limited resources, time, energy, and money? What are the few things that would help you live better and longer if done daily?

The answer lies in the cell.

To properly support your DNA, we must support the cell. Your body is one complex system comprised of more than 37 trillion cells. Cells make energy and need energy.

ENERGY is the currency of life. It allows your body to respond efficiently to all the inputs coming into the system. To work smarter versus harder enables the system to meet every demand with ease leaving you energized instead of drained and depleted.

The cell needs oxygen, nutrients (water, macromolecules of complex carbohydrates, fats, and proteins that break down to micromolecules of vitamins, minerals, and amino acids), and light to properly make and use energy optimally.

The big idea is to take a commonsense approach and begin by reducing the amount, frequency, duration, and exposure to distressing demands. Next, start to pump the brakes to activate the eustress pathways for rest and recovery. Then begin giving your body more of what it needs to increase the production and availability of energy. The more high-octane fuel your system has, the more efficiently it can easily meet life's demands.

Energy is like money. When you have more of it, you can do more with it. Yet, when you spend too much without any income, you quickly go broke, drain your accounts, and end up bankrupt. In terms of your health, this is the fast path to aging, breakdown, and disease.

Your health is your wealth; let's get income coming in so you can heal, thrive, and optimize your life.

Your Epigenetic Longevity Living Blueprint

Focus on the basics first! Do less of what distresses the system. Notice when you feel overloaded so you can pump the brakes by activating your vagus nerve to shift you from overload to ease. This creates micro-moments of eustress that can offer more energy to the system over time.

Once you start to do less of what overloads and more of what energizes, it's time to nourish your cells, so they have the right tools for the job.

Remember, the cell needs: oxygen, nutrients, and light.

Here is how you can get more of these epigenetic longevity lifestyle signals into your body so your cells and DNA can heal and thrive.

Oxygen: You can survive weeks without food, days without water, and only minutes without oxygen. Oxygen is needed in the mitochondria of the cell for oxidative phosphorylation. This is essential to create ATP or usable energy for the body to replicate the DNA, repair, heal, and respond to life's demands.

Today many people are taking in fast, short, and shallow breaths. We are oxygen starved. When you breathe out of your mouth, you quickly dump Carbon Dioxide (CO2). For the cell to get oxygen (O2) in, it needs the CO2 to build up to signal O2 being pushed into use.

Oxygen levels are shown in the research to be tied to our circadian clocks and rhythms, which regulate vital hormones such as cortisol, insulin, leptin, melatonin, and more. When we are low in oxygen, we don't make enough energy, and our circadian rhythms become out of sync, impacting our hormone levels, sleep/wake cycles, and metabolic and immune health, leading to impaired function throughout the body.

To begin using your breath as an epigenetic signal to help slow down the rate of aging and improve energy production, health, and function, consider the following:

Become aware of how you breathe. Fast, short, shallow? Up high in the chest or low in the belly? Is your mouth open or closed?

Choose to optimize your breathing pattern by breathing slow, low, in and out through the nose.

1. See if you can consciously slow your breathing rate without feeling air deprived.
2. Ensure your mouth is closed so you only breathe in and out through the nose.
3. Bring the breath from up high in the chest to low down into the belly so the lungs can expand for greater volume and consumption of O2.

Practice daily.

1. To make it simple, follow the box breath exercise. For example, breathe in through your nose gently for four counts, hold your breath for four counts, breathe out through your nose for four counts, hold for four counts, and repeat.
2. I like to do this before a meal, so I know that at least three times a day, I will consciously box breathe for one to three minutes, helping to improve the main ingredient my body needs to produce high-octane energy.

Nutrients: While nutrition is different for everyone, realize that food is information that talks to your DNA and is used by every cell in your body. What we call nutrigenomics. Here macronutrients of complex carbohydrates, proteins, and fats break down into micronutrients of essential vitamins, minerals, and amino acids. These smaller components are the particles that go into the cell and are used in the various cycles that produce energy, repair, regenerate, replicate, and vitalize the system.

To ensure your cells are getting what they need to slow down the rate and pace of aging, follow these simple macro-balancing principles.

1. Eat three nutrient-dense meals that contain calories from complex carbohydrates, protein, and healthy fats. This can be based on your unique genetics.
2. Drink half of your body weight in ounces daily in clean filtered water. For example, if you weigh 160 lbs., you will want to drink 80 ounces of water daily for proper cell hydration.
3. Consider eating macronutrients that are bioactive nutrients. Bioactive nutrients have actions in the body that may promote good health. This class of foods is a powerful DNA signaler. For a list of epigenetic longevity foods,
4. When you combine macronutrients, your blood sugar is stabilized, giving you more energy for extended periods.
5. For optimal results, eat during the light phase of the day and fast during the dark phase of the day. This syncs you with the natural circadian rhythm for optimal hormonal signaling, energy production, and utilization.

Light: Light and dark signal your cells, hormones, and the processes that create and use energy throughout the system. An area of your brain called the suprachiasmatic nucleus registers light and communicates with every cell in your body to understand where it is in space and time. Each cell has within it an epigenetic clock- yes, your cells tell time. The clocks automatically signal over 600 functions in your body, all based on light and dark, including the regulation and release of key hormones responsible for energy, metabolism, immune function, sexual health, and neurological/ brain function. These epigenetic clocks are linked to the developmental and maintenance processes of biological aging.

The secret to working smarter versus harder is to leverage the natural tools your body uses for greater efficiency. For example, if your body naturally uses light for metabolic function, give it what it needs instead of disrupting or going against that process. When you go against the natural rhythms, you create more work and place more demands on the system.

To use light as an epigenetic longevity lifestyle leaver, consider the following for maximum benefit:

1. Wake up with the light and go outside to get natural sunlight in your eyes for 10 minutes in the morning (ideally before 10 AM) and again in the late afternoon after 3 PM or early evening before sunset.
2. Reduce your exposure to blue light from screens and devices after 5 PM. Change them to amber settings or wear blue blocker glasses.
3. Go dark, ideally two hours before bed. Turn off as many lights in your house as possible and sleep in total darkness with no lights or screens in your room. Light at night disrupts your sleep and hormonal signaling, disrupting your cortisol, glucose, and insulin levels.

Remember, your DNA is always listening. The quality of the inputs impacts the quality of the outputs. So if you want to change any level of your health or life, get curious about the demands on your system and the quality of the inputs and signals coming in. Are they producing the type of vitality, health, and experience you want? If so, keep doing what you are doing. If not, remember this simple epigenetic approach to living better and longer.

Notice if or when you feel overloaded; realize it is a clue, letting you know your DNA is feeling the same way.

Next, choose to lighten your load, nourish your cells, and energize your body and mind so you can thrive by design.

Realize that your DNA is your code of life filled with instructions to help you evolve into greater states of health and flourishing. You are powerful, infinite, and mighty! You don't have to settle for breakdown, sickness, and disease for someone else's idea. When we know better, we can do better, and today, we are sitting at the forefront of a radically new future—one where we understand how to harness our potential.

You, my friend, are filled with innate wisdom and infinite potential. You are writing your epigenetic story. What do you want it to say? You have the choice to do less of what overloads, drains, and depletes you while saying YES, please, and thank you for energizing and filling your soul.

Finally, may you appreciate that a long life lived well is not a sprint. It's the masterpiece of your choices, increasing daily over a lifetime. This is your epigenetic longevity lifestyle stack that creates a long life lived exceptionally and vitally well.

Meet Dr. Rachelle Simpson Sweet

Dr. Rachelle Simpson Sweet was born in the UK and moved to the USA at age 13. She currently lives in Nashua, New Hampshire. Dr. Rachelle has a diverse background, living in many countries and working in various settings. She has interests in finance and real estate and spent many years investing and managing real estate in the US and internationally. She received a formal doctoral education in Clinical Neuropsychology at Nova Southeastern University, a Post-doctoral fellowship at Evanston Northwestern in Pediatric Clinical Neuropsychology, and has now branched into genetic evaluations. She completed a rotation in Health Psychology at Henry Ford Hospital. She holds certifications in Health and Well-being coaching, emphasizing positive psychology and Lifestyle Medicine. Dr. Rachelle has presented research within neuropsychology at professional conferences and contributed writing pieces for three different published books. Having struggled with weight issues and dieting since she was 13 years of age, she understands what it's like to have diets fail and feel like you can't change your body's programs. After years of struggling, observing family members with illness, and studying psychology, neuropsychology, and genetics, she came to understand biological individuality and epigenetics. Dr. Rachelle learned that each of us has a unique blueprint. Based on this, she has developed the B.E.S.T. Living™ coaching program to assist clients in finding their best state of thriving. She also launched the B.E.S.T. Living™ Show podcast.

http://DrRachelleSweet.com, Office@DrRachelleSweet.com

https://drrachellesweet.com
office@DrRachelleSweet.com

ORDER A GENETIC TEST KIT
HTTPS://DRRACHELLESWEET.COM/STORE/

Meet Elizabeth Parrish

Elizabeth (Liz) Parrish, MBA, is the Founder and CEO of BioViva Science USA Inc. BioViva is committed to extending healthy lifespans using gene therapy. BioViva works on combinatorial gene therapies with its proprietary CMV gene therapy delivery platform. Liz is a humanitarian, entrepreneur, author, and innovator. She is a proponent of the Best Choice Medicine plan (BCM), a more efficient and streamlined regulatory model around the use of genetic therapies. She is actively involved in international educational outreach and is a founding member of several 501(c)(3) nonprofits in the space. She considers aging a disease and believes the advancements in this area will lead to healthier and longer lifespans with the added benefit of safer space travel and curative medicine for patients under 20.

Website: Bio VivaSciences.com
info@bioviva-science.com

Meet the Contributing Authors

Abby Kreitler Hand, RN, MSN, FMCHC, is a Preconception Health Consultant and Epigenetic Coach based in beautiful Austin, TX. She specializes in helping women and couples prepare for pregnancy through positive lifestyle change and health optimization. By optimizing the health of both parents before conception, she believes we can build strong families and communities, creating a healthier, more resilient world. Abby is happiest when tromping around the woods with her family and looking at plants.

Website: handwellness.com
Instagram: @handwellness

Corinne and Camille Sullivan H.P are the founders of Liberated Living, Authors, and Co-authors of Epigenetic, Personal Development, and Holistic Psychology eBooks, Audiobooks, Programs, and "Done For You" Internationally Accredited Diploma e-Courses. They offer "DFY' Turnkey Operations and Remote Business Models for Coaches, Practitioners, and Therapists, with Internationally Accredited Certification Courses and an Advanced Signature System in "Genetic Recoding®." Their authored books include Self-Love is the PANACEA, Heal Codependency, and Generational Patterns, "The Powerhouse of the Mind," and more.

Phone: 801-750-9090
Email: contact@liber8edliving.com

Kym Connolly, registered dietitian and epigenetic lifestyle coach, alongside April Wright, are the directors of the company Coabella. Their goal is to empower women to live their higher purpose by coaching them through the health challenges that come with hormonal shifts in menopause. In addition, they focus on slowing down biological aging so women can live lives full of vitality and the freedom that good health brings.

Website: www.dietitiandownunder.com

April Wright is known for combining her talents of writing, online media production, and building websites with her experience and interest in nutrition, epigenetics, and women's health to bring important health information to women all over the world. She's a director of Coabella, a company whose mission is to promote health, menopause, and longevity information to all women.

Website: Coabella.com

Eileen Schutte, CN, FMN, MS, founder, and owner of Unique Nutrition Solutions, offers nutritional and epigenetic counseling for a truly personalized approach for her clients to achieve optimal health. After graduating summa cum laude with her master's degree in human nutrition from the University of Bridgeport, CT, she completed her certification in functional medicine nutrition. In addition, she has completed advanced coursework in nutrigenomics and epigenetics.

Website uniquenutritionsolutions.com
Email: unsnutrition@gmail.com

Self-described as the joyful dietitian, Laurie Kaplan, RDN works to reduce the stress around our food choices to create a beautiful and easy foundation for living. Her goal is to guide clients to calm clarity through a personalized nutrition and lifestyle plan based on their genetic and life goals. My mission is to reduce the world's noise so you can hear the truth of your DNA and truly thrive.

Website: - geneticallynourished.com
Email: geneticallynourished@gmail.com

Melissa Petersen, DC, MS is a visionary female leader in human potential and precision longevity, redefining the limits of human flourishing. She is the founder of the Human Longevity Institute, TEDx Speaker, host of the Human Longevity Podcast, author of the best-selling book, *Codes of Longevity*, an adjunct professor at Life University, and a clinical educator for Medfit. She is on a mission to help people thrive by design, unlock their potential within, and live their longest, healthiest, and most fulfilling lives.

Website: www.docmelissa.com and
https://humanlongevityinstitute.com

Dr. Mickra Hamilton is CEO and Co-Founder of Apeiron Zoh Corporation, a Complex Systems, Precision Performance Ecosystem that Curates Limitless Expression. A Systems Designer and creative disruptor in the field of Precision Human Performance. Dr. Hamilton speaks internationally on the focus areas of the epigenetics of the human system and environment, breathing science, conscious leadership, and peak psychophysiological performance.

A decorated retired Colonel, Hamilton spent 30 years in the USAFR as a Systems Strategist and Human Performance Subject Matter Expert. Leveraging this experience, she works with a complex systems approach for data-driven precision to optimize human and corporate performance and potential.

https://apeironzoh.com/ and https://drmickrahamilton.com/

Dr. Tova Sardot runs Va Sphota Wellness, a human experience optimization practice that uses an eclectic approach to attune complete well-being. Increasing clients' internal coherence on the cellular and energetic levels means evolved states are experienced in life. Tova has a diverse background with a PhD. in Chemistry, M.S. in Complementary Alternative Medicine, B.S. in Physics, and tenure in the industry, government, and academia. This broad experience brings an innovative and cutting-edge perspective from which she works.

Website: www.VaSphotaWellness.com
Email: Tova.Sardot@gmail.com
Phone: 406-282-4838

References

What is B.E.S.T Living™?

Brown, D. J., Arnold, R.; Fletcher, D., and Standage, M. (2017). Human thriving: A conceptual debate and literature review. *European Psychologist, 22*(3), 167–179. https://doi.org/10.1027/1016-9040/a000294

Breathe Chapter

Alaguveni, T.; Devaki, P.R. *Effect of Deep Breathing Exercise on Heart Rate Variability of Different Age Groups,* J Res Med Dent Sci, 2021, 9 (4):267-275.

Battisti-Charbonney, A.; Fisher, J., Duffin, J. *The cerebrovascular response to carbon dioxide in humans.* J Physiol. 2011 Jun 15;589(Pt 12):3039-48. doi: 10.1113/jphysiol.2011.206052. Epub 2011 Apr 26. PMID: 21521758; PMCID: PMC3139085.

Bhasin, M.K.; Denninger, J.W.; Huffman, J.C.; Joseph, M.G.; Niles, H.; Chad-Friedman, E.; Goldman. R.; Buczynski-Kelley, B.; Mahoney, B.A.; Fricchione, G.L.; Dusek, J.A.; Benson, H.; Zusman, R.M.; Libermann, T.A. *Specific Transcriptome Changes Associated with Blood Pressure Reduction in Hypertensive Patients After Relaxation Response* Training. J Altern Complement Med. 2018 May;24(5):486-504. doi: 10.1089/acm.2017.0053. Epub 2018 Apr 4. PMID: 29616846; PMCID: PMC5961875.

Durham, Andrew L. and Adcock, Ian M. *Basic science: Epigenetic programming and the respiratory system,* Breathe Jun 2013, 9 (4) 278-288; doi: 10.1183/20734735.000413.

Howie, H.; Rijal, C.M.; Ressler, K.J. *A review of epigenetic contributions to post-traumatic stress disorder* Dialogues Clin Neurosci. 2019 Dec;21(4):417-428.
doi: 10.31887/DCNS.2019.21.4/kressler. PMID: 31949409; PMCID: PMC6952751.

Jiang, Shui; Postovit, Lynne; Cattaneo, Annamaria; Binder, Elisabeth B.; Aitchison, Katherine J. Epigenetic Modifications in Stress Response Genes Associated with Childhood Trauma, Front. Psychiatry, 08 November 2019Volume 10 – 2019 https://doi.org/10.3389/fpsyt.2019.00808.

McKeown, Patrick. *The Oxygen Advantage: The Simple, Scientifically Proven Breathing Techniques for a Healthier, Slimmer, Faster, and Fitter you,* Publish Year: 2015.

Steffen, P.R.; Austin, T.; DeBarros, A.; Brown, T. *The Impact of Resonance Frequency Breathing on Measures of Heart Rate Variability, Blood Pressure and Mood.* Front Public Health. 2017 Aug 25;5:222. doi: 10.3389/fpubh.2017.00222. PMID: 28890890; PMCID: PMC5575449.

Vlisides, P.E.; Mentz, G.; Leis, A.M.; Colquhoun, D.; McBride, J.; Naik, B.I.; Dunn, L.K.; Aziz, M.F.; Vagnerova, K.; Christensen, C.; Pace, N.L.; Horn, J.; Cummings, K.; Cywinski, J.; Akkermans, A.; Kheterpal, S.; Moore, L.E.; Mashour, G.A. *Carbon Dioxide, Blood Pressure, and Perioperative Stroke: A Retrospective Case-Control Study. Anesthesiology.* 2022 Oct 1;137(4):434-445.
doi: 10.1097/ALN.0000000000004354. PMID: 35960872.

Eat (Epigenetics and Energy) Chapter

Cuthbertson, Lewis, Ph.D. *Can This Gene Influence Your Sugar Eating Habits? (SLC2A2)* August 8th, 2020 – selfdecode.com
https://www.cdc.gov/obesity/data/adult.html

Fawcett, K.A. and Barroso, I. *The genetics of obesity: FTO leads the way.* Trends Genet. 2010;26: 266–274.10.1016/j.tig.2010.02.006.

Heard, E. and Martienssen, R.A. *Transgenerational epigenetic inheritance: myths and mechanisms.* Cell. 2014 Mar 27;157(1):95-109. doi: 10.1016/j.cell.2014.02.045. PMID: 24679529; PMCID: PMC4020004.

Loos, R.J. *The genetics of adiposity.* Curr Opin Genet Dev. 2018 Jun;50:86-95. doi: 10.1016/j.gde.2018.02.009. Epub 2018 Mar 9. PMID: 29529423; PMCID: PMC6089650.

Mehrdad, M.; Doaei, S.; Gholamalizadeh, M., et al. *The association between FTO genotype with macronutrients and calorie intake in overweight adults.* Lipids Health Dis 19, 197 (2020). https://doi.org/10.1186/s12944-020-01372-x

Melhorn, S.J.; Askren, M.K.; Chung, W.K.; Kratz, M.; Bosch, T.A.; Tyagi, V. et al. *FTO genotype impacts food intake and corticolimbic activation.* Am J Clin Nutr. 2018;107: 145–154. 10.1093/ajcn/nqx029.

Mattson, M.P.; Longo, V.D.; Harvie, M. *Impact of intermittent fasting on health and disease processes.* Ageing Res Rev. 2017 Oct;39:46-58. doi: 10.1016/j.arr.2016.10.005. Epub 2016 Oct 31. PMID: 27810402; PMCID: PMC5411330.

National Research Council (US) Subcommittee on the Tenth Edition of the Recommended Dietary Allowances. Recommended Dietary Allowances: 10th Edition. Washington (DC): National Academies Press (US); 1989. 6, Protein and Amino Acids. Available from: https://www.ncbi.nlm.nih.gov/books/NBK234922/

O'Keefe, J.H.; O'Keefe, E.L.; Lavie, C.J. *The Goldilocks Zone for Exercise: Not Too Little, Not Too Much. Mo Med.* 2018 Mar-Apr;115(2):98-105. PMID: 30228692; PMCID: PMC6139866.

Panda, Satchidananda, PhD., Salk Institute for Biological Studies

Ruggiero, C. and Ferrucci, L. *The endeavor of high maintenance homeostasis: resting metabolic rate and the legacy of longevity.* J Gerontol A Biol Sci Med Sci. 2006 May;61(5):466-71. doi: 10.1093/gerona/61.5.466. PMID: 16720742; PMCID: PMC2645618.

Schwartz, M.W.; Seeley, R.J.; Zeltser, L.M.; Drewnowski, A.; Ravussin, E.; Redman, L.M.; Leibel, R.L. *Obesity Pathogenesis: An Endocrine Society Scientific Statement.* Endocr Rev. 2017 Aug 1;38(4):267-296. doi: 10.1210/er.2017-00111. PMID: 28898979; PMCID: PMC5546881.

Su, W.; Huang, J.; Chen, F.; Iacobucci, W.; Mocarski. M.; Dall, T.M.; Perreault, L. *Modeling the clinical and economic implications of obesity using microsimulation.* J Med Econ. 2015;18(11):886-97. doi: 10.3111/13696998.2015.1058805. Epub 2015 Aug 13. PMID: 26057567.

Sleep Chapter

Alhola, P. and Polo-Kantola, P. Sleep deprivation: *Impact on cognitive performance.* Neuropsychiatr Dis Treat. 2007;3(5):553-67. PMID: 19300585; PMCID: PMC2656292.

Beccuti, Guglielmoa,b and Pannain, Silvanaa. *Sleep and obesity. Current Opinion in Clinical Nutrition and Metabolic Care* 14(4):p 402-412, July 2011. doi: 10.1097/MCO.0b013e3283479109.

Carroll, J.E.; Ross, K.M.; Horvath, S.; Okun, M.; Hobel, C.; Rentscher, K.E.; Coussons-Read, M.; Schetter, C.D. *Postpartum sleep loss and accelerated epigenetic aging.* Sleep Health. 2021 Jun;7(3):362-367. doi: 10.1016/j.sleh.2021.02.002. Epub 2021 Apr 24. PMID: 33903077.

McEwen, Bruce S. *In pursuit of resilience: stress, epigenetics, and brain plasticity* -First published: 25 February 2016 https://doi.org/10.1111/nyas.13020

Parsley, Kirk, M.D., *Sleep to Win.*

Puligheddu, M.; Savarese, M.; Spaggiari, M.C.; Simoncini, T. *Italian Association of Sleep Medicine (AIMS) position statement and guideline on the treatment of menopausal sleep disorders.* Maturitas. 2019 Nov;129:30-39. doi: 10.1016/j.maturitas.2019.08.006. Epub 2019 Aug 15. PMID: 31547910.

Reddy, O.C. and van der Werf, Y.D. *The Sleeping Brain: Harnessing the Power of the Glymphatic System through Lifestyle Choices.* Brain Sci. 2020 Nov 17;10(11):868. doi: 10.3390/brainsci10110868. PMID: 33212927; PMCID: PMC7698404.

Silvestri, R.; Aricò, I., Bonanni, E.; Bonsignore, M.; Caretto, M.; Caruso, D.; Di Perri, M.C.; Galletta, S.; Lecca, R.M.; Lombardi, C.; Maestri, M.; Miccoli, M.; Palagini, L.; Provini, F.; Spaeth, A.M.; Dinges, D.F.; Goel, N. *Resting metabolic rate varies by race and by sleep duration.* Obesity (Silver Spring). 2015 Dec;23(12):2349-56. doi: 10.1002/oby.21198. Epub 2015 Nov 5. PMID: 26538305; PMCID: PMC4701627.

Spiegel, K.; Tasali, E.; Penev, P.; Van Cauter, E. *Brief communication: Sleep curtailment in healthy young men is associated with decreased leptin levels, elevated ghrelin levels, and increased hunger and appetite.* Ann Intern Med. 2004 Dec 7;141(11):846-50. doi: 10.7326/0003-4819-141-11-200412070-00008. PMID: 15583226.

Taheri, S.; Lin, L.; Austin, D.; Young, T.; Mignot, E. *Short sleep duration is associated with reduced leptin, elevated ghrelin, and increased body mass index.* PLoS Med. 2004 Dec;1(3):e62. doi: 10.1371/journal.pmed.0010062. Epub 2004 Dec 7. PMID: 15602591; PMCID: PMC535701.

Tian, Y.; Zhao, M.; Chen, Y.; Yang, M.; Wang, Y. *The Underlying Role of the Glymphatic System and Meningeal Lymphatic Vessels in Cerebral Small Vessel Disease. Biomolecules.* 2022; 12(6):748. https://doi.org/10.3390/biom12060748

Westerterp-Plantenga, M.S. *Sleep, circadian rhythm, and body weight: parallel developments.* Proc Nutr Soc. 2016 Nov;75(4):431-439. doi: 10.1017/S0029665116000227. Epub 2016 Apr 27. PMID: 27117840.

Wu, J.C.; Gillin, J.C.; Buchsbaum, M.S.; Schachat, C.; Darnall, L.A.; Keator, D.B.; Fallon, J.H.; Bunney, W.E. *Sleep deprivation PET correlations of Hamilton symptom improvement ratings with changes in relative glucose metabolism in patients with depression.* J Affect Disord. 2008 Apr;107(1-3):181-6.
doi: 10.1016/j.jad.2007.07.030. Epub 2007 Nov 26. PMID: 18031825.

Thrive Chapter

Brown, Daniel J.; Arnold, Rachel; Fletcher, David; Standage, Martyn. "*Human Thriving*." European Psychologist. Published online September 7, 2017, doi:10.1027/1016-9040/a000294.

De Neve, J.E.; Christakis, N.A.; Fowler, J.H.; Frey, B.S. *Genes, Economics, and Happiness.* J Neurosci Psychol Econ. 2012 Nov;5(4):10.1037/a0030292. doi: 10.1037/a0030292. PMID: 24349601; PMCID: PMC3858957.

Guo, Q.; Zheng, R.; Huang, J.; He, M.; Wang, Y.; Guo, Z.; Sun, L.; Chen, P. (2018). *Using Integrative Analysis of DNA Methylation and Gene Expression Data in Multiple Tissue Types to Prioritize Candidate Genes for Drug Development in Obesity. Front.* Genet. 9:663. doi: 10.3389/fgene.2018.00663.

Kanherkar, R.R.; Bhatia-Dey, N.; Csoka, A.B. *Epigenetics across the human lifespan.* Front Cell Dev Biol. 2014 Sep 9;2:49. doi: 10.3389/fcell.2014.00049. PMID: 25364756; PMCID: PMC4207041.

Kanherkar, R.R.; Stair, S.E.; Bhatia-Dey, N.; Mills, P.J.; Chopra, D.; Csoka, A.B. *Epigenetic Mechanisms of Integrative Medicine.* Evid Based Complement Alternat Med. 2017;2017:4365429. doi: 10.1155/2017/4365429. Epub 2017 Feb 21. PMID: 28316635; PMCID: PMC5339524.

Lyubomirsky, S.; Sheldon, K. M.; Schkade, D. (2005). *Pursuing Happiness: The Architecture of Sustainable Change. Review of General Psychology,* 9(2), 111–131. https://doi.org/10.1037/1089-2680.9.2.111

Miller, Kelly https://positivepsychology.com/is-happiness-genetic/

Porath, Christine and Porath, Mike, https://hbr.org/2020/10/how-to-thrive-when-everything-feels-terrible

Epigenetics of Consciousness Chapter

Dawkins, C.R. (2016). *The selfish gene.* Oxford: Oxford University Press. https://www.stagesinternational.com/

Epel, Elissa S., et al., *Accelerated Telomere Shortening in Response to Life Stress."* Proceedings of the National Academy of Sciences 101, no.49 (September 2004): 17312, https://doi.org/10.1073/pnas.0407162101

Furusawa, Chikara and Kaneko, Kunihiko. *"Epigenetic Feedback Regulation Accelerates Adaptation and Evolution."* PLoS ONE 8, no.5 (May 2013): 1, https://doi.org/10.1371/journal/pone.0061251

Low, S.K.; Chin, Y.M.; Ito, H. *et al. Identification of two novel breast cancer loci through large-scale genome-wide association study in the Japanese population. Sci Rep* 9, 17332 (2019). https://doi.org/10.1038/s41598-019-53654-9

Martinez, M.E. (2016), *The mind-body code: How to change the beliefs that limit your health, longevity, and success.* Boulder, Colorado: Sounds True.

Rogier, Eric W., et al. *"Secretory Antibodies In Breast Milk Promote Long-Term Intestinal Homeostasis by Regulating the Gut Microbiota and Host Gene Expressio."* Proceedings of the National Academy of Sciences 111, no. 8 (February 2014): 3074, https://doi.org/10.10.1073/pnas.1315792111

Toth, Miklos. "*Mechanisms of Non-Genetic Inheritance and Psychiatric Disorders, Neuropsychopharmacology*." 40, no.1 (2015): 137, https://doi.org/10.1038/npp.2014.127

Making Change When Change is Hard

https://autoimmune.org/

https://www.cdc.gov/

Genetic Clarity Through Courageous Action My Path to Transformation

White, David. December 2014. *Consolation: The Solace, Nourishment, and Underlying Meaning of Everyday Words*. Canongate Books

How Understanding Epigenetics Helped Me [And Others] Deal With Menopause in a Powerful Way Chapter

Agostini, D.; Zeppa, Donati; S., Lucertini, F.; Annibalini, G.; Gervasi, M.; Ferri Marini C.; Piccoli, G.; Stocchi, V.; Barbieri, E.; Sestili, P. (2018). *Muscle and Bone Health in Postmenopausal Women: Role of Protein and Vitamin D Supplementation Combined with Exercise Training*. Nutrients, 10(8), 1103.
https://doi.org/10.3390/nu10081103

Ahn, J.K.; Hwang, J.; Lee, M.Y.; Kang, M.; Hwang, J.; Koh, E.M.; Cha, H.S. (2021). *How much does fat mass change affect serum uric acid levels among apparently clinically healthy Korean men?* Therapeutic advances in musculoskeletal disease, 13, 1759720X21993253. https://doi.org/10.1177/1759720X21993253

Almeida, M.; Soares, M.; Fonseca-Moutinho, J.; Ramalhinho, A. C.; Breitenfeld, L. (2021). *Influence of Estrogenic Metabolic Pathway Genes Polymorphisms on Postmenopausal Breast Cancer Risk*. Pharmaceuticals (Basel, Switzerland), 14(2), 94. https://doi.org/10.3390/ph14020094

Baker, James M.; Al-Nakkash, Layla; Herbst-Kralovetz; Melissa M. (2017). *Estrogen–gut microbiome axis: Physiological and clinical implications*, Maturitas, Volume 103, Pages 45-53, https://doi.org/10.1016/j.maturitas.2017.06.025

Bermingham. K.; Linenberg, I.; Hall, W., et al. 2022. *Menopause is associated with postprandial metabolism, metabolic health, and lifestyle: The ZOE PREDICT study*. eBioMedicine, Volume 85, 104303, https://doi.org/10.1016/j.ebiom.2022.104303

Berry, Sarah; Mazidi, Mohsen; Franks, Paul; Valdes, Ana; Sattar, Naveed; Wolf, Jonathan; Hadjigeorgiou, George; David, Drew; Chan, Andrew; Segata, Nicola; Asnicar, Francesco; Spector, Tim; Hall, Wendy. *Impact of Postprandial Lipemia and Glycemia on Inflammatory Factors in over 1000 Individuals in the US and UK: Insights from the PREDICT 1 and InterCardio Studies*, Current Developments in Nutrition, Volume 4, Issue Supplement_2, June 2020, Page 1518, https://doi.org/10.1093/cdn/nzaa068_003

Berry, Sarah; Wyatt, Patrick; Franks, Paul; Blundell, John; O'Driscoll, Ruairi; Wolf, Jonathan; Hadjigeorgiou, George; Drew, David; Chan, Andrew; Spector, Tim; Valdes, Ana. *Effect of Postprandial Glucose Dips on Hunger and Energy Intake in 1102 Subjects in US and UK: The PREDICT 1 Study*, Curren Developments in Nutrition, Volume 4, Issue Supplement_2, June 2020, Page 1611, https://doi.org/10.1093/cdn/nzaa063_009

Cho, S.K.; Winkler, C.A.; Lee, S-J; Chang, Y.; Ryu, S., The Prevalence of Hyperuricemia Sharply, *Increases from the Late Menopausal Transition Stage in Middle-Aged Women*. Journal of Clinical Medicine. 2019; 8(3):296. https://doi.org/10.3390/jcm8030296

Colton, Carol A.; Brown, Candice M.; Vitek, Michael P. 2005. *Sex steroids, APOE genotype, and the innate immune system*. Neurobiology of Aging 26 (2005) 363–372 Fernando Lizcano and Guillermo Guzmán.

Estrogen Deficiency and the Origin of Obesity during Menopause (2014). BioMed Research International Volume 2014, Article ID 757461, http://dx.doi.org/10.1155/2014/757461

Figtree, G.A.; Noonan, J.E.; Bhindi, R.; Collins, P. *Estrogen receptor polymorphisms: significance to human physiology, disease, and therapy.* Recent Pat DNA Gene Seq. 2009;3(3):164-71. doi: 10.2174/187221509789318397. PMID: 19673701.

Fischer, L.M.; da Costa, K.A.; Kwock, L.; Galanko, J.; Zeisel, S.H. *Dietary choline requirements of women: effects of estrogen and genetic variation.* Am J Clin Nutr. 2010 Nov;92(5):1113-9. doi: 10.3945/ajcn.2010.30064. Epub 2010 Sep 22. PMID: 20861172; PMCID: PMC2954445.

Ghosh, S. and Klein, R.S. *Sex Drives Dimorphic Immune Responses to Viral Infections.* J Immunol. 2017 Mar 1;198(5):1782-1790. doi: 10.4049/jimmunol.1601166. PMID: 28223406; PMCID: PMC5325721.

Gordon, J.L.; Rubinow, D.R; Eisenlohr-Moul, T.A.; Leserman, J.; Girdler, S.S.; *Estradiol variability, stressful life events, and the emergence of depressive symptomatology during the menopausal transition.* Menopause. 2016 Mar;23(3):257-66. doi: 10.1097/GME.0000000000000528. PMID: 26529616; PMCID: PMC4764412.

Hak, A. E. and Choi, H.K. (2008). *Menopause, postmenopausal hormone use and serum uric acid levels in US women*--the Third National Health and Nutrition Examination Survey. Arthritis research and therapy, 10(5), R116. https://doi.org/10.1186/ar2519

Hewings-Martin, Y. and Giordano, F. 2022. *Diet may counteract menopause metabolism change, Zoe's study shows.* Online blog accessed 4 Dec 2022. https://joinzoe.com/learn/menopause-metabolism-study https://www.nice.org.uk/guidance/ng23/ifp/chapter/managing-your-symptoms accessed 25 Jan 2022

Hu, L.; Ji, J.; Li, D., et al. *The combined effect of Vitamin K and calcium on bone mineral density in humans: a meta-analysis of randomized controlled trials.* J Orthop Surg Res16, 592 (2021) https://doi.org/10.1186/s13018-021-02728-4

Jacobs, Emily G.; Weiss, Blair K.; Makris, Nikos Makris; Whitfield-Gabrieli, Sue; Buka, Stephen L.; Klibanski, Anne; Goldstein, Jill M. (2016). *Impact of Sex and Menopausal Status on Episodic Memory Circuitry in Early Midlife,* Journal of Neuroscience vol 36 (39) 10163-10173; doi:10.1523/JNEUROSCI.0951-16.2016.

Johnson, R.J., Stenvinkel, P., Martin, S.L., Jani, A., Sánchez-Lozada, L.G., Hill, J.O. and Lanaspa, M.A. (2013), *Redefining metabolic syndrome as a fat storage condition based on studies of comparative physiology.* Obesity, 21: 659-664. https://doi.org/10.1002/oby.20026

Jones, N., Kiely, J.; Suraci, B.; Collins, D.J.; de Lorenzo, D.; Pickering, C.; Grimaldi, K.A. *A genetic-based algorithm for personalized resistance training.* Biol Sport. 2016 Jun;33(2):117-26. doi: 10.5604/20831862.1198210. Epub 2016 Apr 1. PMID: 27274104; PMCID: PMC4885623.

Kaneki, M. *Genomic approaches to bone and joint diseases. New insights into molecular mechanisms underlying protective effects of Vitamin K on bone health.* Clin Calcium. 2008 Feb;18(2):224-32. Japanese. PMID: 1824589.

Karvinen, S.; Juppi, H.K.; Le, G.; Cabelka, C.A.; Mader, T.L.; Lowe, D.A.; Laakkonen, E.K. *Estradiol deficiency and skeletal muscle apoptosis: Possible contribution of microRNAs.* Exp Gerontol. 2021 Feb 4;147:111267. doi: 10.1016/j.exger.2021.111267. Online ahead of print. PMID: 33548486

Kheirouri, Sorayya and Alizadeh, Mohammad. (2021). *MIND diet and cognitive performance in older adults: a systematic review,* Critical Reviews in Food Science and Nutrition, doi: 10.1080/10408398.2021.1925220.

Krause, W.C.; Rodriguez, R.; Gegenhuber, B., et al. *Oestrogen engages brain MC4R signaling to drive physical activity in female mice.* Nature 599, 131–135 (2021). https://doi.org/10.1038/s41586-021-04010-3

Li, F.; Boon, A.C.M.; Michelson, A.P. et al. *Estrogen hormone is an essential sex factor inhibiting inflammation and immune response in COVID-19.* Sci Rep 12, 9462 (2022). https://doi.org/10.1038/s41598-022-13585-4

Link, R. 2021 7 *Healthy Foods That Are High in Hyaluronic Acid,* online https://www.healthline.com/nutrition/hyaluronic-acid-diet, accessed 5 Dec 2022.

Lyssenko, V.; Nagorny, C.L.; Erdos, M.R.; Wierup, N.; Jonsson, A.; Spégel, P.; Bugliani, M.; Saxena, R.; Fex, M.; Pulizzi, N.; Isomaa, B.; Tuomi, T.; Nilsson, P.; Kuusisto, J.; Tuomilehto, J.; Boehnke, M.; Altshuler, D.; Sundler, F.; Eriksson, J.G.; Jackson, A.U.; Laakso, M.; Marchetti, P.; Watanabe, R.M.; Mulder, H.; Groop, L. *Common variant in MTNR1B associated with increased risk of type 2 diabetes and impaired early insulin secretion.* Nat Genet. 2009 Jan;41(1):82-8. : 10.1038/ng.288. Epub 2008 Dec 7. PMID: 19060908; PMCID: PMC3725650.

Malacara, J.M.; Pérez-Luque, E.L.; Martínez-Garza, S.; Sánchez-Marín, F.J. *The relationship of estrogen receptor-alpha polymorphism with symptoms and other characteristics in post-menopausal women.* Maturitas. 2004 Oct 15;49(2):163-9. DOI: 10.1016/j.maturitas.2004.01.002 PMID: 15474761.

Nappi, R. E.; Martella, S.; Albani, F.; Cassani, C., Martini, E.; Landoni, F. (2022). *Hyaluronic Acid: A Valid Therapeutic Option for Early Management of Genitourinary Syndrome of Menopause in Cancer Survivors?* Healthcare (Basel, Switzerland), 10(8), 1528. https://doi.org/10.3390/healthcare10081528

Newson, L.; Manyonda, I.; Lewis, R. et al. 2021. *Sensitive to Infection but Strong in Defense—Female Sex and the Power of Oestradiol in the*

COVID-19 Pandemic. Front. Glob. Women's Health, Sec. Sex and Gender Differences in Disease https://doi.org/10.3389/fgwh.2021.651752

Pakharenko, L. *Effect of estrogen receptor gene ESR1 polymorphism on development of the premenstrual syndrome.* Georgian Med News. 2014 Oct;(235):37-41. PMID: 25416214.

Parry, B.L. *Towards improving recognition and management of perimenopausal depression.* Menopause. 2020 Apr;27(4):377-379. doi: 10.1097/GME. 0000000000001519. PMID: 32108733

Piroddi, M.; Albini, A.; Fabiani, R.; Giovannelli, L.; Luceri, C.; Natella, F.; Rosignoli, P.; Rossi, T.; Taticchi, A.; Servili, M.; Galli, F. *Nutrigenomics of extra-virgin olive oil: A review.* Biofactors. 2017 Jan 2;43(1):17-41. doi: 10.1002/biof.1318. Epub 2016 Sep 1. PMID: 27580701.

Prasad, Megha; Matteson, Eric L.; Herrmann, Joerg; Gulati, Rajiv; Rihal, Charanjit S., Lerman, Lilach O.; Amir Lerman. 2016. *Uric Acid Is Associated With Inflammation, Coronary Microvascular Dysfunction, and Adverse Outcomes in Postmenopausal Women.* Hypertension. 2017;69:236–242 https://www.ahajournals.org/doi/abs/10.1161/HYPERTENSIONAHA.116.08436

Qiu, Juanjuan, MMDa; Du, Zhenggui, MDa,b,c; Liu, Jingping, MDd; Zhou, Yi MDd; Liang, Faqing MMDa,b; Lü, Qing MDa,b, *A prospective case–control study.* Medicine: November 2018 – Volume 97 - Issue 47 - p e13337 DOI: 10.1097/MD.0000000000013337.

Rajoria, S.; Suriano, R.; Parmar, P. S.; Wilson, Y. L.; Megwalu, U.; Moscatello, A.; Bradlow, H. L.; Sepkovic, D.W.; Geliebter, J.; Schantz, S.P.; Tiwari, R.K. (2011). *3,3'-diindolylmethane modulates estrogen metabolism in patients with thyroid proliferative disease: a pilot study.* Thyroid: official journal of the American Thyroid Association, 21(3), 299–304. https://doi.org/10.1089/thy.2010.0245

Sánchez-Villegas, A.; Pérez-Cornago, A.; Zazpe, I. et al. *Micronutrient intake adequacy and depression risk in the SUN cohort study.* Eur J Nutr 57, 2409–2419 (2018). https://doi.org/10.1007/s00394-017-1514-z

Sandell, M.A. and Collado, M. *Genetic variation in the TAS2R38 taste receptor contributes to the oral microbiota in North and South European locations: a pilot study.* Genes Nutr 13, 30 (2018). https://doi.org/10.1186/s12263-018-0617-3

Sandell, M.; Hoppu, U.; Mikkilä, V.; Mononen, N.; Kähönen, M.; Männistö, S.; Rönnemaa, T.; Viikari, J.; Lehtimäki, T.; Raitakari, O.T. (2014). *Genetic variation in the hTAS2R38 taste receptor and food consumption among Finnish adults.* Genes and nutrition, 9(6), 433. https://doi.org/10.1007/s12263-014-0433-3

Satyanarayana, R.; Pondugula, Patrick C.; Flannery, Kodye L.; Abbott, Coleman, Elaine S.; Mani, Sridhar; Temesgen, Samuel; Xie, Wen. 2015. *Diindolylmethane, a naturally occurring compound, induces CYP3A4 and MDR1 gene expression by activating human PXR,* Toxicology Letters, Volume 232, Issue 3, Pages 580-589, https://doi.org/10.1016/j.toxlet.2014.12.015

Schuit, S.C.E.; de Jong, F.H.; Stolk, L.; Koek, W.N.H.; van Meurs, J.B.J.; Schoofs, M.W.C.J.; Zillikens, M.C.; Hofman, A.; van Leeuwen, J.P.T.M.; Pols, H.A.P.; Uitterlinden, A.G. (2005) *Estrogen receptor alpha gene polymorphisms are associated with estradiol levels in postmenopausal women,* European Journal of Endocrinology eur j endocrinol,153(2), 327-334. Retrieved Dec 4, 2022, from https://eje.bioscientifica.com/view/journals/eje/153/2/1530327.xml

Shivapriya, Manchali; Kotamballi, N.; Chidambara, Murthy; Bhimanagouda, S. Patil. *Crucial facts about health benefits of popular cruciferous vegetables,* 2012. Journal of Functional Foods, Volume 4, Issue 1, Pages 94-106, https://doi.org/10.1016/j.jff.2011.08.004

Van Ballegooijen, A.J.; Pilz, S.; Tomaschitz, A.; Grübler, M.R.; Verheyen, N. *The Synergistic Interplay between Vitamins D and K for Bone and Cardiovascular Health:* A Narrative Review. Int J Endocrinol. 2017;2017:7454376. doi: 10.1155/2017/7454376. Epub 2017 Sep 12. PMID: 29138634; PMCID: PMC5613455.

Van der Veen, J.N.; Kennelly, J.P.; Wan, S.; Vance, J.E.; Vance, D.E.; Jacobs, R.L. *The critical role of phosphatidylcholine and phosphatidylethanolamine metabolism in health and disease.* Biochim Biophys Acta Biomembr. 2017 Sep;1859(9 Pt B):1558-1572. doi: 10.1016/j.bbamem.2017.04.006. Epub 2017 Apr 11. PMID: 28411170.

Weber, M.T., Rubin, L.H., Maki, P.M. *Cognition in perimenopause: the effect of the transition stage.* Menopause. 2013 May;20(5):511-7. doi: 10.1097/gme.0b013e31827655e5 PMID: 23615642; PMCID: PMC3620712

Yang, X.; Guo, Y.; He, J.; Zhang, F.; Sun, X.; Yang, S.; and Dong, H. (2017). *Estrogen and estrogen receptors in the modulation of gastrointestinal epithelial secretion.* Oncotarget, 8(57), 97683–97692 https://doi.org/10.18632/oncotarget.18313

Yoo, Seung-Schik; Gujar, Ninad; Hu, Peter; Jolesz, Ferenc A.; Walker, Matthew P. (2007). *The human emotional brain without sleep — a prefrontal amygdala disconnect,* Current Biology, Volume 17, Issue 20, Pages R877-R878.

The Many Genes of Histamine Intolerance - Genomics was a Game Changer for Me Chapter

Barone, M.; D'Amico, F.P.; Brigidi, Patrizia; Turroni, S. *Gut microbiome-micronutrient interaction: The key to controlling the bioavailability of minerals and vitamins?* Biofactors, Wiley, 2022:48:307-314, Feb. 23, 2022.

Comas-Baste, Oriol, et al., *Histamine Intolerance: The Current State of the Art,* Biomolecules, Aug. 14, 2020.

Fokerts, J., et al., *Effect of Dietary Fiber and Metabolites on Mast Cell Activation and Mast Cell-Associated Diseases,* Frontiers in Immunology, 9:1067, May 29, 2018

Schnedl, Wolfgang J. and Enko, Dietmar. *Histamine Intolerance Originates in The Gut,* Nutrients 2021, 13, 1262, April 12, 2021.

How a Biophysics Perspective on Epigenetics Can Change Your Entire Life! Chapter

Becker, Robert O, *The Body Electric: Electromagnetism and the Foundation of Life,* William Morrow, 2020.

Cao-Lei, Lei; Laplante, David P.; King, Suzanne. *"Prenatal maternal stress and epigenetics: Review of the human research."* Current Molecular Biology Reports 2.1 (2016): 1625.

Chae, K.S.; Kim, S.C.; Kwon, H.J., et al. *Human magnetic sense is mediated by a light and magnetic field resonance-dependent mechanism.* Sci Rep 12, 8997 (2022)

Church, Dawson. *Genie in Your Genes: Epigenetic Medicine and the New Biology of Intention,* Hay House, September 2018.

Crum, Alia J. and Langer, Ellen J. 2007. *Mind-set matters: Exercise and the placebo effect.* Psychological Science 18, no. 2: 165-17.

Dossey, Larry. *"Prayer and medical science: a commentary on the prayer study by Harris et al. and a response to critics."* Archives of Internal Medicine 160.12 (2000): 1735-1738.

Fetterman, Jessica L. and Ballinger, Scott W. *Mitochondrial genetics regulate nuclear gene expression through metabolites,* 2019, 116 (32) 15763-157.

Gaudi, Simona, et al. *"Epigenetic mechanisms and associated brain circuits in the regulation of positive emotions: A role for transposable elements."* Journal of Comparative Neurology 524.15 (2016): 2944-2954.

Gouin, J.P.; Kiecolt-Glaser, J.K., et al. *The influence of anger expression on wound healing.* Brain Behav Immun. 2008 Jul;22(5):699-708.

Ho, Mae-Wan; Popp, Fritz-Albert; Ulrich Warnke. *Bioelectrodynamics and biocommunication.* World Scientific, 1994.

Ironson, Gail; Stuetzle, Rick; Fletcher, Mary Ann. *"An increase in religiousness/spirituality occurs after HIV diagnosis and predicts slower disease progression over four years in people with HIV."* Journal of general internal Medicine 21.5 (2006): S62-S68.

LeDoux, Joseph. *Rethinking the Emotional Brain,* Neuron, Volume 73, Issue 4, 2012, Pages 653-676.

Marcus, Gary. *The Birth of the Mind: How a Tiny Number of Genes Creates the Complexity of Human Thought,* Basic Books, 2014.

McCrady, R., Children, D. The Appreciative Heart: The psychophysiology of positive emotions and optimal functioning, Institute of HeartMath, Boulder Creek, CA; 2002.

Popp, F.A.; Nagl, W.; Li, K.H.; Scholz, W.; Weingärtner, O.; Wolf, R. Biophoton emission. *New evidence for coherence and DNA as the source,* Cell Biophys, 1984 Mar;6(1):33-52.

Rossi, E. (2002). *The psychobiology of gene expression* (p. 36). New York, NY., Norton.

Wang, Connie X., Hilburn, Isaac A. Hilburn, et al. *Transduction of the Geomagnetic Field as Evidenced from Alpha-band Activity in the Human Brain,* eNeuro 18 March 2019, ENEURO.0483-18.2019.

The Hidden Powerhouse of The Mind Chapter

Carlson, E. B, and Dalenberg C. J. (2000), *A conceptual framework for the impact of traumatic experiences. trauma, Violence, and Abuse,* 1(1), 4-28. doi: 10.1177/1524838000001001002.

Jablonka, E., and Raz, G. (2009). *Transgenerational epigenetic inheritance: prevalence, mechanisms, and implications for the study of heredity and evolution.* Q Rev Biol, 84(2), 131-176. doi: 10.1086/598822.

Psychology, Wiki-Fandom, 2022. Mere-exposure effect. Retrieved from: https://psychology.fandom.com/wiki/Mere-exposure_effect on 10-06-22

Raghunathan. R, (2012). *Familiarity Breeds Enjoyment.* Psychology Today. Retrieved from:
https://www.psychologytoday.com/us/blog/sapient-nature/201201/familiarity-breeds-enjoyment

Rodgers et al., 2013; Gapp et al., 2014; Pang et al., 2017; *Epigenetic inheritance is, in certain cases, reduced primarily to paternal contribution,* e.g., Yeshurun and Hannan, 2018).

Rodgers, et al., 2013; Gapp, et al., 2014; Pang, et al.,(2017); Yeshurun and Hannan, 2018. *Epigenetic inheritance* retrieved from:
https://www.frontiersin.org/articles/10.3389/fnmol.2018.00292/full

Van Otterdijk, Sanne D. and Michels, Karen B. 2016; Pang et al., (2017) *Epigenetic inheritance refers to the transmission of certain epigenetic marks to offspring.*

The Epigenetics of Longevity Chapter

Adamovich, Y.; Ladeuix, B.; Golik, M.; Koeners, M.P. Asher, G. *Rhythmic Oxygen Levels Reset Circadian Clocks through HIF1α.* Cell Metab. 2017 Jan 10;25(1):93-101. doi: 10.1016/j.cmet.2016.09.014. Epub 2016 Oct 20. PMID: 27773695.

Breit, S.; Kupferberg, A.; Rogler, G.; Hasler, G. *Vagus Nerve as Modulator of the Brain-Gut Axis in Psychiatric and Inflammatory Disorders.* Front Psychiatry. 2018 Mar 13;9:44. doi: 10.3389/fpsyt.2018.00044. PMID: 29593576; PMCID: PMC5859128.

Horvath, Steve and Raj, Kenneth. *DNA methylation-based biomarkers and the epigenetic clock theory of aging.* Nat Rev Genet 2018 Jun;19(6):371-384.

Passarino, G. and De Rango, F. Montesanto A. *Human longevity: Genetics or Lifestyle? It takes two to tango.* Immun Ageing. 2016 Apr 5;13:12. doi: 10.1186/s12979-016-0066-z. PMID: 27053941; PMCID: PMC4822264.

Serin, Yeliz. and Acar, Tek N. *Effect of Circadian Rhythm on Metabolic Processes, and the Regulation of Energy Balance.* Ann Nutr Metab 2019;74:322-330. doi: 10.1159/000500071.

Weinhold B. *Epigenetics: the science of change.* Environ Health Perspect. 2006 Mar;114(3):A160-7. doi: 10.1289/ehp.114-a160. PMID: 16507447; PMCID: PMC1392256.

What is Stress? National Research Council (US) Committee on Recognition and Alleviation of Distress in Laboratory Animals. Recognition and Alleviation of Distress in Laboratory Animals. Washington (DC): National Academies Press (US); 2008. 2, Stress and Distress: Definitions. Available from: https://www.ncbi.nlm.nih.gov/books/NBK4027/

Young, Sergey. The Science and Technology of Growing Young. BenBella Books, 2021.

Zeng, Yi and Shen, Ke. *Resilience significantly contributes to exceptional longevity.* Curr Gerontol Geriatr Res. 2010;2010:525693. doi: 10.1155/2010/525693. Epub 2010 Dec 6. PMID: 21197075; PMCID: PMC3004383.

Additional Reading

The Biology of Belief: Unleashing the Power of Consciousness, Matter, and Miracle by Bruce Lipton

Change Your Genes, Change Your Life: Creating Optimal Health with the New Science of Epigenetics by Kenneth R. Pelletier

Codes of Longevity by Melissa Petersen

The Comfort Crisis by Michael Easter

The Disease Delusion: Conquering the Cause of Chronic Illness for a Healthier, Longer, and Happier Life By Jeffrey S. Bland

Epigenetics, How Environment Shapes Our Genes by Richard C Francis

Epigenetic Revolution by Nessa Cary

The Genomic Kitchen by Amanda Archibald, RD

Organumics, An Epigenetic Re-Framing of Consciousness, Life, and Evolution by Ben Callif

Oxygen Advantage by Patrick G. Mckeown

The Surrender Experiment by Michael Singer

Synchrodestiny by Deepak Chopra

Made in United States
North Haven, CT
17 March 2023

34220302R10141